Parent-Teacher Guide for

Ray's™ New Arithmetics

Parent-Teacher Guide for

Ray's™ New Arithmetics

A guide for teaching from these books in Ray's Arithmetic Series:

- Primary
- Intellectual
- Practical
- Higher

By Ruth Beechick, Ed.D.

Parent-Teacher Guide for Ray's New Arithmetics

For information about other Mott Media publications visit our website at www.mottmedia.com.

Printed in the United States of America

ISBN-10: 0-88062-071-4
ISBN-13: 978-0-88062-071-0

Contents

Introduction

page

What Are Ray's Arithmetics?

Joseph Ray was born in 1807 on a farm about ten miles from the farm on which William McGuffey had been born just seven years earlier. This coincidence foreshadowed the similar careers of these two giants of textbook writing. What McGuffey did for readers, Ray did for arithmetics.

Ray's popular arithmetics saw over one thousand editions and sold perhaps 120 million copies, more than any other arithmetic in America in the 1800s. Ray's books were known for treating comprehensively all important fundamentals of arithmetic. They progressed systematically from simple to complex and included many puzzle-like problems to challenge competent students. They stated principles and rules and followed them with examples. They emphasized mental arithmetic to precede written arithmetic as a means to assure understanding. These and other features won for Ray's a wide acclaim and use for nearly a century of American school life.

Ray's arithmetics were also known for their low cost. Our forebears were concerned about cost just as we are, so much so that publishers sometimes joined two books in one binding so parents would have only one book to buy. In 1837 three of Ray's books sold for 10¢, 20¢, and 35¢, respectively. While not matching that price, Mott Media has succeeded in bringing to today's parents and teachers a reprinting of Ray's classic arithmetics, low in cost and complete in scope. Students who master these books are well prepared for any testing and higher education they aspire to in today's world.

Christians sometimes ask, "Does the study of arithmetic glorify God?" The answer is yes. While learning arithmetic, children develop their God-given, natural, Biblical mode of thinking. Biblical thinking begins with the premises that God created everything and that basic truths of the creation are self-evident to us. We know, for instance, that we are in time and space. The concept of number—with unity and plurality—seems natural to us. So do the concepts of motion, equality, causation, and order.

Pagan humanist reasoning denies that outer reality is truth. Truth is derived in the minds of humans, and if human logic cannot prove something, it cannot be called a truth. Thus, since we cannot prove by our frail logic systems that there is a God, God is not truth.

But in the study of arithmetic, even pagans still use a Biblical mode of thinking. The basic, unprovable truths are acknowledged by all, and the study of arithmetic is built upon them. With this mode of thinking and study, it is natural to view God's creation as orderly. Stars keep time more perfectly than clocks can ever manage, crystals teach solid geometry, musical tones and overtones vibrate in mathematical patterns that man discovers rather than creates. The heavens declare to us the glory of God and the firmament shows His handiwork.

Yes, children can learn arithmetic to the glory of God.

Ray's Arithmetic Series

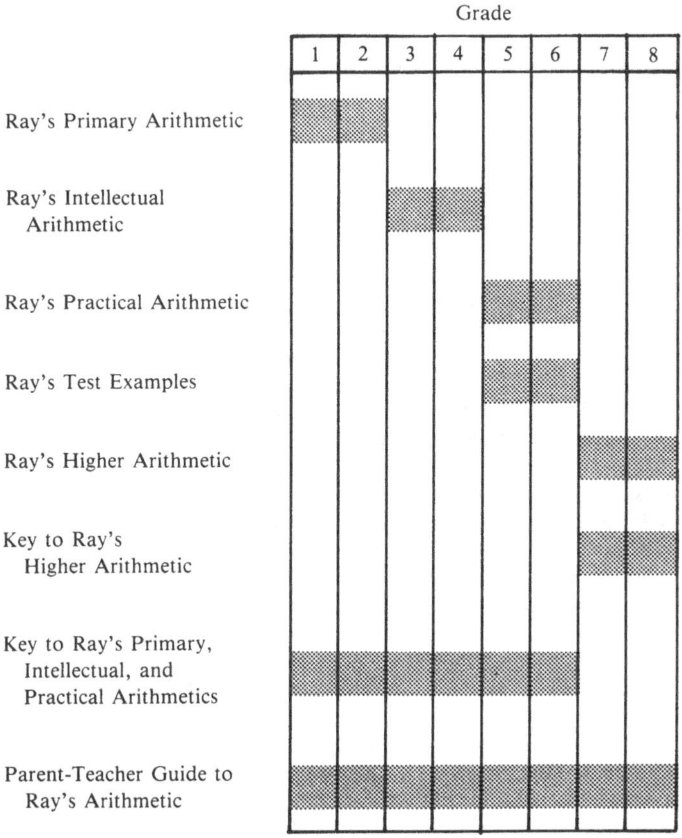

This chart shows suggested grade levels for using the books in Ray's arithmetic series. The books themselves are not numbered, so it is easy to place children with special needs for either remedial or advanced work.

Scope and Sequence

Grade 1

Notation. Read and write figures to 100.

Addition. Add numbers with sums up to 20, and solve problems using this skill.

Subtraction. Subtract numbers with sums up to 19 and solve problems using this skill.

Counting. Count objects in one-to-one correspondence; count forward and backward by 2s, 3s, etc.

Measurement. Gain practical experience with calendars, clocks, rulers, and other measuring devices; use money; know simple shapes and simple fractions.

Vocabulary. All number names to 100.

Grade 2

Review. Review and reteach as necessary all first grade learning on notation, addition, and subtraction.

Multiplication. Memorize multiplication tables to 5s or to 10s, and solve problems of multiplication.

Division. Memorize division tables to 5s or to 10s, and solve problems of division.

Signs. Learn equals, plus, minus, multiplication, and division signs; read and write problems using them.

Measurement. Gain experience with money, weights, and measures.

Vocabulary: plus, minus, multiplied by, divided by, equals.

Grade 3

Notation. Read and write whole numbers, fractions, mixed numbers, and signs for plus, minus, multiplied by, divided by, and equals.

Addition. Memorize facts to 12s; mentally solve problems with higher numbers, involving more than two addends; counting by 3s, 4s, etc.

Subtraction. Memorize facts to 12s; mentally solve problems with higher numbers and combining other operations with subtraction.

Multiplication. Memorize facts to 12s; solve problems which involve only multiplication, or combine it with other operations.

Division. Memorize facts to 12s; solve division problems and problems with any combination of operations.

Fractions. Read, write, and compute with simple fractions.

Vocabulary: sum, equals, minuend, subtrahend, difference, remainder, minus, multiplicand, multiplier, product, divisor, dividend, quotient.

Grade 4

Basic Operations. Review addition, subtraction, multiplication, and division with whole numbers.

Fractions. Add, subtract, multiply, and divide fractions and mixed numbers.

Measurement. Know U.S. money, liquid and dry measures, measures of weight, length, time.

Ratio. Compute and solve problems with whole and mixed numbers.

Percentage. State fractional parts as hundredths; apply to interest, commission and other business uses.

Vocabulary: denominator, numerator, proper fraction, improper fraction, lowest terms, (least) common denominator, ratio, percent, commission, premium, interest, principle, discount.

Grade 5

Notation. Read, write, and use large numbers in the Arabic system; understand the arrangement of orders and periods; use Roman numbers I, V, X, L, C, D, and M.

Basic Operations. Add with large numbers and with carrying; subtract with large numbers and with borrowing; multiply with methods; divide with short and long methods.

Measurement. Compute with simple and compound numbers in all kinds of measurements and weights; reduce compound numbers to convert, such as feet to inches or longitude to hours; find areas and volumes.

Factors. Find factors and use them in many kinds of computations.

Vocabulary: addends, integer, abstract number, concrete number, simple number, compound number, degrees (in a circle), factor(ing), multiple, prime number, composite number, prime factor, (least) common divisior, (least) common multiple.

Grade 6

Notation. Read and write decimals, understanding the orders; use parentheses, brackets, ratio and proportion signs, exponents, index numbers, and radical signs.

Decimals and Percentage. Compute in all four operations; estimate; solve problems of interest, discount, exchange, insurance, and taxation.

Ratio and Proportion. Find missing terms and compute values.

Powers and Roots. Raise numbers to any power, extract square and cube roots.

Measurement. Convert metric measures; and compute in metric measures of length, area, capacity, and weight.

Geometry. Find lengths and areas of common plane shapes and volumes of common solids.

Vocabulary: average, mean, base, rate, percentage, antecedent, consequent, involution, evolution, power, exponent, square root, cube root, radical sign, hypotenuse, perpendicular, acute angle, obtuse angle, plane, polygon, quadrilateral, trapezium, trapezoid, parallelogram, rhombus, hexagon, octagon, circumference, diameter, pyramid, sphere, and others.

Grade 7

Philosophical Understandings. Learn to see arithmetic as both a science and an art, understanding the scope of arithmetic with its operations and understanding the place of principles, rules, proofs, and other important ideas in the field of arithmetic.

Numeration and Basic Operations. Use these with proficiency, and understand more fully the principles involved in them.

Properties of Numbers. Understand more fully the kinds of numbers: whole, fractional, mixed, abstract, concrete, prime, composite, even, odd, perfect and imperfect. And understand principles and rules in dealing with the various kinds.

Common and Decimal Fractions. Gain proficiency in all the operations with fractions. Understand their properties. Learn to handle circulating decimals.

Measurements. Understand the units and compute in all common measurements, including metric. Gain proficiency in handling reductions, aliquot parts, and other aspects of compound numbers.

Ratio and Proportions. Gain understanding and proficiency in ratio and simple and compound proportions.

Grade 8

Percentage. Learn principles for finding bases, rates, and percentages. Apply the knowledge to practical uses as in interest and discounts.

Business Math. Learn about capital, profit and loss, partnerships, bankruptcy and other business topics and computations.

Alligation. Find proportional parts for business or science purposes.

Involution and Evolution. Compute powers and roots.

Series. Understand arithmetical and geometrical progressions, and make computations involving them.

Geometry. Compute lengths, areas and volumes.

> Problem solving is an inseparable feature of every topic. It is a major strand running through all grades.

Tips for Teaching Ray's Arithmetics

Two Main Principles

Do you have Arithmetic Anxiety? If you answer Yes, you are not alone, but are among a large number of moderns who suffer from this malady. But the cure is now in your hands. You are about to teach children by a process that will prevent them from developing Arithmetic Anxiety. And while you are teaching, your own case will be healed.

Is it going to be easy? Not if your idea of teaching arithmetic is to say, "Shut up, sit down, and do that page of problems." Ray's arithmetic does not work that way at all. It is not a recess for the teacher.

In Ray's system, the teacher actually teaches every inch of the way. You must lead your children through mental and oral work. In the early years, especially, there is almost no writing. But there is plenty of thinking. This feature of Ray's is extremely important to the success of young children in arithmetic. Remember, it's their brains they are going to take to college later on, not their papers. So don't be in a hurry to have them fill papers with rows of problems. Follow Ray's system faithfully and you will see excellent results.

Ray's textbooks are organized according to the logic of arithmetic, so it is easy for you, and for your pupils too, to know exactly what you are studying and when you have learned it. The books build from simple to complex, they state rules clearly, they frequently review.

And Ray seems to have known principles of child development that we think we are just discovering in our time. He made use of the principles by starting with manipulative objects and mental work, and waiting until later to introduce abstract symbols, such as division signs. This feature in itself goes a long way toward eliminating Arithmetic Anxiety. And it does far more. It builds a firm foundation of understanding that lasts through all the later difficult topics.

INSTANT REPLAY

Features of Ray's Arithmetic

1) *Organized according to logic.* It presents arithmetic topics in orderly fashion, moving from simple to complex, and stating rules clearly.

2) *Presented according to principles of child development.* It reviews frequently and it moves in order through the stages of arithmetic growth in children.

Three Growth Stages

What are the stages of arithmetic growth in children? There are three major ones: 1) the manipulative stage, 2) the mental image stage, and 3) the abstract stage. Since you have been through these yourself, you don't need to be a psychologist to understand them. But you must take time to familiarize yourself with them. Each stage is described below. If you read the descriptions now, and again occasionally after you start teaching, it will open your understanding to much that is going on in your child's mind. Knowing these stages will help you solve many teaching problems that may come your way. Here are the three stages. Learn them well.

1) Manipulative Stage

Young children learn a lot of arithmetic from the real world around them. A child with one block can get another block and see that he now has two. A parent's or teacher's first job here is to teach him the words *one* and *two*. It is too early to teach the abstract symbols $1 + 1 = 2$.

In the early years of arithmetic study, a large proportion of the work should be with manipulative objects. The children themselves will give you clues that they are ready to do problems in their heads.

Do not rush the transition. Use the time in this stage to provide a rich variety of number experiences. Work the problems in *Ray's Primary Arithmetic* orally, using objects to show what happens. Have a shelf full of homemade and commercial games that involve numbers. And use everyday experiences—setting four forks on the table, choosing two loaves of bread at the store, and so forth. During this stage, the children can be learning to read and write the numerals, just as they learn to read and write letters of the alphabet.

This manipulative stage corresponds in general with the stage Piaget has called preoperational. By that term, he meant that children do not do operations in their heads, but are learning to do them outside their heads—with objects to manipulate.

When older children begin a new topic, they, also, should be allowed time to manipulate objects before they must work problems in their heads. For instance, when beginning fractions, let them cut an orange into parts or divide a group into parts. After enough experience of this kind they can move easily into the mental and written work.

This is so powerful a learning principle, that you can use it to help your children through practically any learning problem in arithmetic. At a trouble spot, involve the child in meaningful experiences at the manipulative level. Help him move objects or make drawings to think through what really happens. After enough experience at this level, he will internalize the thinking and become able to do the work in his head.

2) Mental Image Stage

As the child gains experience with real objects, he gains skill in making mental images and thinking with them. He finds this a faster way to work out problems, so he naturally drops the use of objects. There is no need for you to force this transition.

At this stage, too, a large proportion of the work should be oral. If the written forms of problems are introduced too early, there may be undue emphasis on where to write what numerals, and the child is distracted from the far more important basic understandings.

This stage corresponds in general to what Piaget has called the concrete operational stage. Piaget's choice of words has been somewhat unfortunate in America, because Americans were used to thinking of concrete operations as those done with objects outside the head, whereas Piaget referred to internalized images that children manipulate inside their heads. The term "mental image stage" eliminates the confusion.

3) Abstract Stage

Arithmetic study is full of abstractions. The numerals themselves are abstract symbols which stand for something in the real world. The signs are more abstract yet, since they tell us what to do with the abstract numerals. And ideas such as percents, fractions, ratios, and so forth are well up the abstraction ladder.

Adults think with abstractions so routinely that sometimes we forget what the child is up against when he meets abstractions in his school work. Remember, the child is slowly growing in his ability to handle abstractions. Teach patiently and, when necessary, meaningfully involve the child in activities at the mental image level, or even the manipulative level, to build up the understandings a child needs. You want your child to know how to write computations. For instance, you will teach him where the figures all go in a long division problem, but writing it is not necessarily understanding it. Continue giving the child many arithmetic experiences and let him talk about his work and explain it. And over time, you will see him grow in ability to handle abstractions.

This stage corresponds in general to the stage Piaget has called the abstract operational stage. At first, children can use only simple abstractions in their heads, and no one has found a good way to speed up the process. It takes both time and practice to move on to more difficult abstractions.

INSTANT REPLAY

Three Growth Stages

1) *Manipulative Stage.* The child works problems with real objects.
2) *Mental Image Stage.* The child internalizes the thinking and manipulates objects in his head.
3) *Abstract Stage.* The child can begin thinking about abstract ideas and can make use of symbols in his thinking.

How Much Drill?

Your answer to this question is important for several reasons. On one hand, you need to require enough drill that your children will master each concept. And on the other hand, you need to move on to new concepts rapidly enough that your children remain challenged and interested. In between, is an area of efficient teaching and learning.

Ray's arithmetics are arranged with each new concept introduced in logical order. Following each concept are problems which give the children practice and drill on the concept, and you are free to decide how many of the problems to use with any class or any particular student. When you feel your children have mastered a concept, move on. Don't be a slave to the book; let it be your servant, instead.

If you are tutoring one or a few children and doing much of the work orally and mentally, you will easily know when the children are ready for something new. Trust your judgment; it will serve you better than tests in most cases. If you are teaching classroom groups, the unit tests will help to supplement your judgment. Or you can easily make tests by selecting problems from the unit being studied.

On basic number facts, it is a good idea to continue drill longer than on other skills. The basic facts are the multiplication and division tables and the addition and subtraction combinations. These all should be overlearned. That is, drill should continue even after it seems the children know the facts. The reason, of course, is that accuracy and speed on the basic facts help the children do better in every other arithmetic procedure.

So, as teacher, you must find ways to keep the children interested in extra drill on basic facts. Some ideas in the game section of this manual will help you with this job. Simple races help, too. Try speed tests, with each child racing against his own best time.

When children are young and in the early grades of arithmetic skills, are they better off without formal teaching? Many parents and teachers believe so. If you are one of these, you need not use the kinds of activities that are usually thought of as drill. Instead, you can provide many arithmetic experiences in natural settings around the home or workshop or schoolroom. You can study the text and the scope and sequence chart to see what concepts and understandings to build, and work toward those in an informal manner.

How much drill? You hold the answer for the children you teach. And you need not be afraid of making a wrong decision. The worst that can happen is that you will later decide to back up and reteach something. The best that can happen is that your classroom becomes a more efficient learning center than ever before.

INSTANT REPLAY

Drill

1) Use more or less drill according to your children's needs.

2) On basic number facts, drill for overlearning, accuracy, and speed.

3) With young children, follow your school's or your personal philosophy concerning informal learning versus formal teaching and drill.

Teaching First Grade

Using *Ray's New Primary Arithmetic*

page

Teaching First Grade

The main objectives of the first grade curriculum are to learn to read, write, and count with numbers under 100; and to add and subtract numbers with sums up to 20. This covers Lessons I through XXXVII of *Ray's Primary Arithmetic*, and the planning guide shows how you may complete this in a 36-week school year. Children, of course, vary in the speed with which they catch on to arithmetic concepts, and as teacher you must adjust the lessons to the children. This is more important at first grade than anywhere else, because you want the children to have a happy, successful start in arithmetic study.

Ray's system relies heavily on teacher guided discovery learning. This kind of learning has a long history of success, and it is a Biblical principle. The Ethiopian eunuch knew he couldn't learn from the book of Isaiah unless someone would guide him, and Philip did (Acts 8:31). Jesus said that the Holy Spirit would "guide" us into all truth (John 16:13). Experienced teachers and recent researches agree with this principle of guiding, and they acknowledge it to be the most effective method for getting good results.

Ray's also begins computation exclusively with mental and oral work. With this kind of start, research shows, children develop patterns of thinking that are efficient and that make use of quite elegant mathematical concepts, such as commutativity. For instance, if a child must add 2 and 8, he soon learns that it is more efficient to start with 8 and add on 2. This is a greater achievement than it might first appear to us adults. The child, in doing this, shows that he also understands the concept of cardinality—the ordering of numbers—and that he realizes he can begin counting anywhere, not only at 1. But if children have been taught specific ways of writing problems and certain ways of working them, this kind of development is slowed.

So for more excellent results, you should faithfully follow the mental system in Ray's arithmetics. Avoid the temptation to hurry the children into writing rows and rows of problems so you can put them to work and take a recess. Many children are better off not starting formal arithmetic study quite so early as our society tends to think. So, particularly in a home school setting or small one-room school, you can use real-life situations, manipulative objects, and games to help children develop in their understanding of numbers. Relax and take things at the children's pace.

Become familiar with the three stages of arithmetic growth and other helps in this manual. And look over the following typical schedules and the planning guide. Then you are on your way to a great year with your first graders.

A Typical Day

There probably is no really typical day in first grade arithmetic because there is such a variety of things to do and children vary so much. But you do need a plan.

Spend the first part of the period on the main work for the day. Determine this from the planning guide and the weekly schedule. For instance, on Mondays you will usually work directly from the textbook on the new lesson for the week. Wind up this work before the children tire from concentrating. Then for the rest of the period, play number games with your children, or let them play with each other. Or sometimes do a project, such as measuring ingredients and mixing up a recipe. A number of projects are suggested in the project section of this manual, and they will suggest others to you. This practical approach to numbers gives children a strong foundation of number understanding.

A Typical Week's Schedule

Monday. Use the textbook lesson orally as an informal diagnosis of the children's learning needs.

Tuesday. Use real objects, flashcards, games, chalkboard, slates, or paper to practice on the new content of the lesson. When the children write, in most cases it should be answers only, and not the full problems.

Wednesday. Try the textbook lesson again. Notice improvement, and diagnose any trouble spots.

Thursday. Repeat practice as on Tuesday. Or if the lesson is mastered, use this day for work on one of the measurement topics or other enrichment ideas in this manual.

Friday. Review the textbook lesson.

On Thursdays and other opportune occasions, you can teach the practical topics of time, measurement, and others listed below. Use these objectives several times during the year as a checklist to see what topics the children need more experience with.

Calendars. Read days of the week, dates of the month, and months of the year. Use calendars to look backward and to plan ahead.

Clocks. Read time on digital clocks. Read to half hours on hand clocks.

Rulers. Use rulers and measure items to the nearest inch and to the nearest centimeter.

Shapes. Recognize circles, triangles, squares, and other rectangles.

Money. Count and use pennies, nickels, dimes.

Fractions. Recognize halves, quarters, and thirds of an item. Find one-half of a group of items.

Planning Guide for Grade 1

(Ray's Primary Arithmetic)

UNIT: NUMBERS AND FIGURES

Week 1

Lesson I

Objective: to count up to 10 objects and to express their numbers by figures.

Practice counting numerous real objects around the classroom or home. Write numerals on chalkboard and paper. Play domino- and lotto-type games. This lesson must be thoroughly mastered before moving to Lesson II.

Week 2

Lesson II

Objective: to count up to 40 objects.

Count and play games as in Lesson 1. Children who have mastered Lesson 1 may extend their writing of figures up to 20.

Week 3

Lesson III

Objective: to count up to 70 objects.

Review numbers 1 to 20. These must be mastered. Work on counting higher numbers according to the ability of the children. Learn number rhymes and songs. Play number games. (See the game section of this manual.)

Week 4

Lesson IV

Objective: to count up to 100 objects.

Review numbers learned so far and work on higher counting according to the ability of the children.

Week 5

Lesson V

Objective: to be able to read figures from 1 to 100.

This lesson gives a new kind of practice on the numbers. Follow the suggested weekly schedule closely. Use flashcards. Where else can children find numbers to read? (Book pages, calendars, ads, and so forth.)

Week 6

Lesson VI

Objective: to write figures from 1 to 100.

This lesson winds up the first unit. For children who have difficulty, concentrate the practice on figures 1 to 20. These must be mastered before proceeding to the next unit. Advanced pupils may enjoy writing numbers higher than 100.

UNIT: ORAL EXERCISES

Week 7

Lesson VII

Objective: to understand and manipulate numbers of items up to 5.

Use real objects and do the problems orally. Sometimes use fruit or nuts and let the children eat the treats after their lesson. This work with concrete objects will build a strong understanding of arithmetic principles. Do not be in a hurry to have the children write the problems in figures or to introduce the abstract symbols of plus, minus, and equal signs.

Week 8

Lesson VIII

Objective: to understand and manipulate numbers of objects up to 8.

Continue the oral work with real objects. For variety, let the children bring treats or other objects to use. Also, let them think up problems to ask each other.

Review some of the number writing work from the first unit.

Week 9

Lesson IX

Objective: to understand and manipulate numbers of objects up to 10.

Continue the oral work as in Lesson VIII. Review number writing.

UNIT: ADDITION

Week 10

Lesson X

Objective: to add by counting items in pictures.

From concrete objects, the lessons now move to pictures of objects. Find more pictures and make up additional problems. Let the children make up problems.

Week 11

Lesson XI

Objective: to add 1 to all numbers up to 10.

The previous work with real objects and pictured objects should help the children have mental images of the addition problems in this unit. For children who have difficulty, you can "act out" these problems with real fruit and real pennies.

Flashcards at this stage can say "1 and 1," "2 and 1," and so forth, with answers on the reverse side. Children should be able to give the answers quickly.

Week 12

Lesson XII

Objective: to add 2 to all numbers up to 10.

Use the story problems orally. Can the children make up problems? Drill on the combinations with 2 until the children can answer rapidly.

Week 13

Lesson XIII

Objective: to add 3 to all numbers up to 10.

Use story problems and drill as in Lesson XII. Provide plenty of practice with concrete items for those who need it.

Week 14

Lesson XIV

Objective: to add 4 to all numbers up to 10.

Continue as in previous addition lessons, using oral practice. Beginning with this lesson, you can adjust the objective for children who have difficulty with the work. Require sums only up to 10 for them. In other words, require them to add 4 to all numbers up to 6. Also, keep reviewing previous lessons.

Week 15

Lesson XV

Objective: to add 5 to all numbers up to 10.

For children who have difficulty with these, teach only up to 5 and 5, or as far as the children are able to go.

Week 16

Lesson XVI

Objective: to add 6 to all numbers up to 10.

Adjust the objective as needed for your pupils. Some may not need much practice with concrete objects anymore. When you work orally with children, it is easy to see who needs it and who doesn't. Let that part of the practice fall off naturally as children are able to do the problems without the help of manipulative objects.

Week 17

Lesson XVII

Objective: to add 7 to all numbers up to 10.

Again, adjust the objective as needed. For children who are not learning these easily, give lots of experience with number games and lots of review on easier addition problems.

Week 18

Lesson XVIII

Objective: to add 8 to all numbers up to 10.

Continue as in the previous lessons of this unit. All the work is still oral. For written work, you can review Lessons I to IV. Try to get everyone to master the writing of numbers from 1 to 20. Most children can learn to write from 1 to 100, and a few children may go beyond 100.

Week 19

Lesson XIX

Objective: to add 9 to all numbers up to 10.

Continue as in previous lessons of this unit.

Week 20

Lesson XX

Objective: to add 10 to all numbers up to 10.

Continue as in previous lessons of this unit.

Weeks 21 to 23

Lessons XXI to XXIII

Objective: to review and master additional facts up to sums of 20.

These three lessons provide time for reviewing and overlearning the addition facts. This would be a good time to update pupil progress records. Let children work in pairs to teach each other and master the facts they haven't learned yet. Lesson XXI introduces problems with three addends. Lesson XXII introduces a variety of counting problems. And Lesson XXIII uses all the skills in story problems. Children who have thoroughly mastered the work thus far have an excellent foundation for all later arithmetic skills. If your children can read and write numbers only up to 10 and can do sums up to 10, you may still let them move on to the subtraction unit and work with the lower numbers there.

UNIT: SUBTRACTION

Week 24

Lesson XXIV

Objective: to subtract by counting objects in pictures.

The picture in this lesson helps children to see mental images of what happens in subtraction. Find more pictures and make up problems about them, too. Help your children to use the vocabulary of subtraction. They should readily use phrases like "4 from 5," "6 less 1," or "6 take away 1." It is not necessary to learn the word *minus* at this time; children can learn it later when the minus sign is introduced.

Week 25

Lesson XXV

Objective: with sums up to 10, to subtract when 1 is the subtrahend or the remainder.

Let children manipulate objects to work out problems for as long as they need to. When the children's minds can internalize the manipulations they will drop the practice of using objects. Let each child follow his own developmental time table.

Week 26

Lesson XXVI

Objective: with sums up to 11, to subtract when 2 is the subtrahend or the remainder.

Throughout this subtraction unit you may adjust the objective, and make a basic course by concentrating on sums of 10 and less. Remember that these lessons are still to be done mentally and orally. Read again the suggested weekly plan and use either it or your own adjusted version to teach the lessons of this subtraction unit.

Weeks 27 to 34

Lessons XXVII to XXXIV

Objectives: with sums up to 19, to subtract when numbers 3, 4, 5, 6, 7, 8, and 9 are the subtrahend or the remainder.

Proceed with each lesson as in the previous subtraction lessons. For a basic course you can adjust the objective each week, aiming for mastery of at least the sums up to 10. For an advanced course, provide extra challenge and enrichment from the section in this manual on games and projects.

Week 35

Lessons XXXV and XXXVI

Objectives: to review and master subtraction facts up to sums of 19, and to review addition facts.

During this week you can evaluate each of your pupils and update their progress records. The review lessons provide time for any reteaching that may be needed, and for mastery and overlearning, and for testing. The best way to test primary children is by observing as they manipulate objects and by questioning them orally. With *Ray's Primary Arithmetic*, you have been doing that daily, so you should have a good idea of the achievement level of each child. You can use these review weeks for individual remedial work in addition to your class reviews from the text.

Week 36

Lesson XXXVII

Objective: to be able to use subtraction facts in backwards counting and in life story problems.

The counting problems give a new dimension to understanding subtraction. They also lay a basis for study of division later on. Repeat these until your children can do them easily. Let the children think up other counting problems. They don't have to end with zero. Also, let the children think up life story problems.

PROGRESS RECORD

Grade 1

	count objects		read and write numbers		add		substract		works mostly at		
Names	to 20	to 100	to 20	to 100	sums to 10	sums to 20	sums to 10	sums to 19	manipu-lative stage	mental-image stage	Comments

Testing First Graders

In the primary grades, formal tests are not very reliable as a way to measure achievement. Besides that, written tests cannot evaluate many of the important areas of children's knowledge. You, as teacher, can get a good evaluation almost daily as you lead the children in oral work and keep simple records of what they can and cannot do.

Nevertheless, there are times when you may want to test by some means. One common use of testing is for placement, to determine where a child or a group needs to be in the textbook. Another is for diagnosis, to see in what areas a child needs more teaching and practice. Still another use is at the end of a unit as a final check-up to see if the children have learned the material well enough to move on to the next unit.

Tests for each unit are given on the following pages of this manual. At first grade level you will obtain the best results by individual oral testing. But if you find it necessary to test groups of children, you may have them write answers after you read aloud the problems. You have the publisher's permission to make copies of the tests if you are using Ray's arithmetic series.

Test Schedule

Unit 1: Numbers and Figures. Use the two tests, "Counting Objects" and "Writing Figures."

These may be used as pretests to help you decide how much time to spend on the work in this unit. And after the unit, they may be used again as posttests. On the counting test use real objects for the child, as you follow the test sheet yourself. With the figures test, it may be easier to have the children write on blank sheets of paper. But if you use reproduced test sheets, supervise closely enough that you can help the children write their figures in the proper spaces.

Unit 2: Oral Exercises. Use the test, "Solving Problems, Concrete Level."

It is important that this test be given with real objects or, at the very least, with pictures. If your children do not function well at this concrete level, give them lots of practice in this before you expect them to do problems in their heads. You may move on to the addition and subtraction units, but do most of the work with real objects or an abacus or counters of some kind.

Unit 3: Addition. Use the two tests, "Addition Facts" and "Solving Addition Problems."

For the problems test, let the children use objects or counters if they are at the concrete level. Or let them work in their heads if they are at the mental-image level. Note on their progress charts which way they are functioning. For the facts test, the answers may be memorized.

Test for the lower sums for children in the basic course, and the higher sums for children in the advanced course.

Unit 4: Subtraction. Use the two tests, "Subtraction Facts" and "Solving Subtraction Problems."

As in the addition unit, let the children work either at the concrete level or at the mental-image level. And test for lower numbers or higher numbers according to the ability of the children.

Counting Objects

Name ______________________ Date ____________

How many?

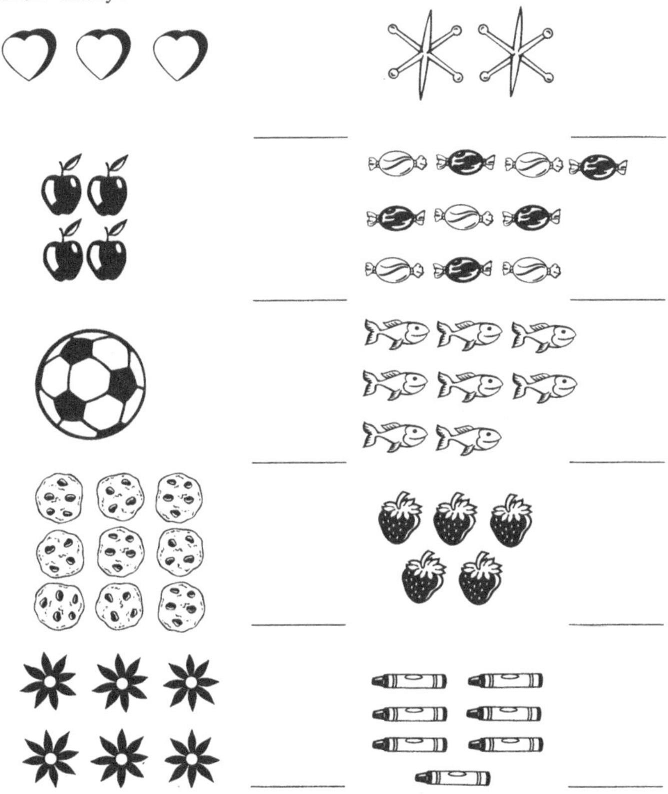

Score __________

Writing Numbers

Name ______________________________ Date _______________

To ten:

________________	________________	________________
three	seven	five

To twenty:

________________	________________	________________
eighteen	fourteen	eleven

To one hundred:

________________	________________	________________
sixty-eight	ninety	thirty-three
________________	________________	________________
twenty-one	fifty	forty-two

Score ____________

Solving Problems

(Concrete-Level Oral Test)

Name ______________________________ Date _______________

1. How many are 2 cents and 2 cents?

2. I have three pieces of candy. If I give you 2 of them, how many will I have left?

3. Here are 3 puzzle pieces. If you put 2 of them in the puzzle, how many will be left?

4. Here are 6 jacks. How many twos can you pick up?

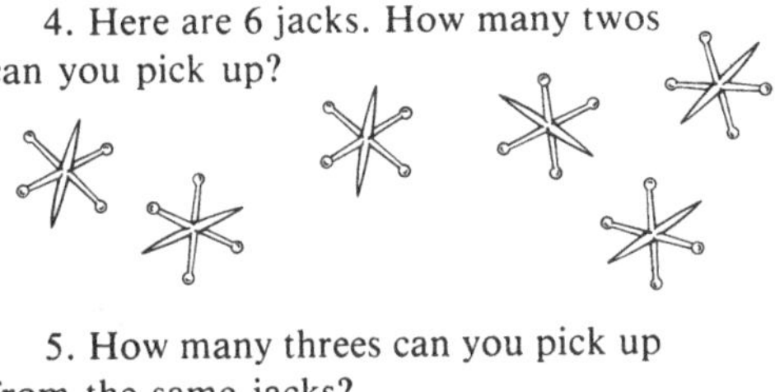

5. How many threes can you pick up from the same jacks?

6. Children are playing with 8 toy cars. If Mark takes 1 car home, how many will be left?

7. Sara picked 4 apples, Adam picked 4 and Jane picked 2. How many apples will be in the basket?

8. Every day for four days, Carl put a penny in his bank. How many pennies did he save?

9. If you save 2 pennies each day for four days, how many pennies will you have?

10. There are 8 cookies. How many should each twin get?

Score ___________

Addition Facts

Name ______________________ Date ______________

To 5s:

2 and 5 are ________	3 and 4 are ________
5 and 2 are ________	1 and 5 are ________
4 and 4 are ________	3 and 2 are ________
1 and 1 are ________	5 and 5 are ________
2 and 1 are ________	2 and 4 are ________
4 and 5 are ________	2 and 2 are ________
3 and 3 are ________	4 and 1 are ________
3 and 1 are ________	4 and 3 are ________

To 10s:

7 and 7 are ________	9 and 4 are ________
10 and 6 are ________	9 and 5 are ________
8 and 6 are ________	2 and 10 are ________
6 and 7 are ________	9 and 9 are ________
10 and 10 are ________	2 and 6 are ________
7 and 9 are ________	4 and 7 are ________
9 and 1 are ________	9 and 2 are ________
9 and 5 are ________	3 and 10 are ________
6 and 5 are ________	6 and 1 are ________
10 and 4 are ________	8 and 2 are ________
6 and 4 are ________	7 and 3 are ________
1 and 7 are ________	4 and 8 are ________
6 and 9 are ________	7 and 10 are ________
9 and 10 are ________	9 and 8 are ________
7 and 2 are ________	7 and 8 are ________
8 and 10 are ________	6 and 3 are ________
6 and 6 are ________	5 and 10 are ________
8 and 8 are ________	8 and 3 are ________

Score ________

Solving Addition Problems

Name ______________________ Date ____________

Lower Numbers

1. Henry had 1 peach and his mother gave him 2 more. How many had he then? ________

2. John had 2 cupcakes and his mother gave him 5 more. How many had he then? ________

3. Mary had 3 books and her brother Oliver had 4. How many had they both? ________

4. How many are 4 cents and 4 cents? ________

5. There are 5 letters in my last name and 4 letters in my first name. How many letters are in both names? ________

Score __________

Higher Numbers

6. There are 6 pigs in one pen and 5 pigs in another pen. How many pigs are in both pens? ________

7. There are 7 boys on one bench and 4 boys on another bench. How many boys are there? ________

8. Eight geese flew onto a pond and 5 more geese came a moment later. How many geese were on the pond? ________

9. Joseph caught 9 fish in one stream and 4 in another. How many fish did he catch? ________

10. Sheila paid 10 dollars for a doll and 3 dollars for a game. How many dollars did she spend? ________

Score __________

Subtraction Facts

Name ______________________________ Date _______________

To 5:

1 from 5 leaves ______	3 from 4 leaves ______
2 from 3 leaves ______	4 from 5 leaves ______
1 from 4 leaves ______	1 from 2 leaves ______
4 from 4 leaves ______	2 from 4 leaves ______
1 from 3 leaves ______	3 from 3 leaves ______
2 from 5 leaves ______	3 from 5 leaves ______
2 from 2 leaves ______	1 from 3 leaves ______

To 10:

6 from 10 leaves ______	2 from 9 leaves ______
2 from 8 leaves ______	9 from 10 leaves ______
4 from 7 leaves ______	3 from 8 leaves ______
6 from 6 leaves ______	2 from 6 leaves ______
8 from 8 leaves ______	4 from 6 leaves ______
3 from 10 leaves ______	8 from 10 leaves ______
1 from 9 leaves ______	7 from 9 leaves ______
2 from 10 leaves ______	5 from 10 leaves ______
2 from 7 leaves ______	4 from 8 leaves ______
4 from 10 leaves ______	3 from 9 leaves ______
3 from 7 leaves ______	7 from 10 leaves ______
1 from 10 leaves ______	1 from 7 leaves ______
7 from 8 leaves ______	6 from 8 leaves ______
1 from 6 leaves ______	8 from 9 leaves ______
5 from 6 leaves ______	5 from 9 leaves ______
4 from 9 leaves ______	6 from 7 leaves ______
5 from 7 leaves ______	5 from 8 leaves ______
6 from 9 leaves ______	3 from 6 leaves ______

Score ____________

Solving Subtraction Problems

Name ______________________________ Date _______________

Lower Subtrahends or Remainders

1. Alice had 8 chickens and one was killed. How many had she then? ________

2. John had 8 marbles but he lost 2 of them. How many did he have left? __________

3. Six people were in a car. Three of them got out. How many were still in the car? __________

4. Eliza had 6 birds in a cage and she let 2 of them out. How many remained in the cage? __________

5. Eight sailing boats went out on the lake, and 5 came back before a storm. How many were still out? __________

Score ____________

Higher Subtrahends or Remainders

6. James wrote 7 numbers on a slate and then erased 1 of them. How many numbers remained? __________

7. Ella bought an orange for 7 cents and sold it for 11 cents. How much did she gain? __________

8. If 16 children are in a room and 8 of them leave, how many will remain? __________

9. Thomas had 12 walnuts and ate 9 of them. How many nuts did he have left? __________

10. Someone owed me 12 dollars, and he repaid all but 2 dollars. How much money did he pay me? __________

Score ____________

Answer Key

COUNTING OBJECTS. 3, 2, 4, 10, 1, 8, 9, 5, 6, 7.

WRITING NUMBERS. 3, 7, 5, 18, 14, 11, 68, 90, 33, 21, 50, 42.

SOLVING PROBLEMS. **1)** 4 cents. **2)** 1. **3)** 1. **4)** 3. **5)** 2. **6)** 7. **7)** 10. **8)** 4. **9)** 8. **10)** 4.

ADDITION FACTS. To 5s, Column 1: 7, 7, 8, 2, 9, 6, 4. **Column 2:** 7, 6, 5, 10, 6, 4, 5, 7. **To 10s, Column 1:** 14, 16, 14, 13, 20, 16, 10, 14, 11, 14, 10, 8, 15, 19, 9, 18, 12, 16. **Column 2:** 13, 14, 12, 18, 8, 11, 11, 13, 7, 10, 10, 12, 17, 17, 15, 9, 15, 11.

SOLVING ADDITION PROBLEMS. Lower Numbers: 1) 3. **2)** 7. **3)** 7. **4)** 8. **5)** 9. **Higher Numbers. 6)** 11. **7)** 11. **8)** 13. **9)** 13. **10)** 13.

SUBTRACTION FACTS. To 5, Column 1: 4, 1, 3, 0 , 2, 3, 0. **Column 2:** 1, 1, 1, 2, 0, 2, 2. **To 10, Column 1:** 4, 6, 3, 0, 0, 7, 8, 8, 5, 6, 4, 9, 1, 5, 1, 5, 2, 3. **Column 2:** 7, 1, 5, 4, 2, 2, 2, 5, 4, 6, 3, 6, 2, 1, 4, 1, 3, 3.

SOLVING SUBTRACTION PROBLEMS. Lower: 1) 7. **2)** 6. **3)** 3. **4)** 4. **5)** 3. **Higher: 6)** 6. **7)** 4. **8)** 8. **9)** 3. **10)** 10.

Teaching Second Grade

Using *Ray's New Primary Arithmetic*

page

Teaching Second Grade

The main objective of the second grade curriculum is to achieve ability in the four basic operations of arithmetic—adding, subtracting, multiplying, and dividing. It covers Lessons XXXVIII through LXXXIX in *Ray's Primary Arithmetic.*

To accomplish this, you should begin the year by reviewing the first grade units of adding and subtracting. Use Ray's method of working orally with the children. You can ask a question, then let the children figure out the answer either with manipulative materials or with mental images. Do not require that they write down the problems, although sometimes you may ask them to write the answer only, to give practice in notation.

Children trained in this system move easily into mental multiplication and division because these operations are simply extensions of addition and subtraction. That is, a child who can add 2 and 2 and 2 has no trouble figuring out what three 2s are. He multiplies as easily as he adds. Difficulty arises for many children when they are taught too soon our symbolic ways of writing down the various operations. One reason for this is that they have not yet reached the abstract stage of thinking. The idea of multiplication is abstract, but doing it is not. For instance, a child playing jacks can see what two fives are. Or he might easily count up what four nickels are worth. But if he first meets these problems abstractly, as ______ x 5 = 10, or 2 x 5 = ______, it is more difficult to understand.

So keep your children doing the problems according to Ray's mental system. Late in the year, in the unit on signs, they may try writing out the problems. It will be easy after they have achieved facility in the operations.

Notice in the planning guide that after week 7 there are two branches. At that time of year, make the best decision you can about which branch to take your children on. But your decision can be changed later. For instance, if you start on the advanced course but find after two or three weeks that the larger numbers are difficult for your children, you can back up and take them on the basic course instead. Or if your children master the basic course you can move to the advanced course and take up some of the additional work there.

A Typical Day

Arithmetic should be a morning study, coming either first on the schedule or second, after penmanship. Spend the first part of the period with the main work for the day. This may be a new textbook lesson, or a reteaching of something difficult from the previous day, or a drill on memorizing basic number facts. Wind up this lesson before the children tire from concentrating. Then for the rest of the period, assign a project such as those suggested in the project section of this manual. Or play some number games together. Or let the children play among themselves. These more relaxed number activities are important in building good understanding of numbers.

A Typical Week's Schedule

Monday. Use the textbook lesson orally as an informal diagnosis of the children's learning needs.

Tuesday. Use real objects, flashcards, games, chalkboard, slates, or paper to practice on the content of the new lesson.

Wednesday. Try the textbook lesson again. Notice improvement, and diagnose any trouble spots. Or if the week's plan covers more than one lesson, start a new lesson today.

Thursday. Repeat practice as on Tuesday. Or if the lesson is mastered, use this day for work on one of the measurement topics or other enrichment ideas in this manual.

Friday. Review the textbook lesson. Make notes for yourself about anything you want to provide additional practice on next week. On review units, you may need to take new lessons on Thursday or Friday in order to keep the schedule.

Planning Guide for Grade 2

(Ray's Primary Arithmetic)

Weeks 1 to 3

Lessons I to XL

Objective: to review and relearn the facts, skills, and concepts studied in first grade.

In the vacation between first grade and second grade there seems to be more forgetting of skills than at any other level. Second graders profit greatly from a good review. An average class might follow this schedule, but take more time if your group needs it.

Week 1. Review the unit of numbers and figures, which includes Lessons I to VI.

Week 2. Review the unit on addition, Lessons X to XXIII.

Week 3. Review the unit on subtraction, Lessons XXIV to XXXVIII.

UNIT: MULTIPLICATION

Week 4

Lessons XXXVIII to XL

Objectives: to understand the idea of multiplication, and to master the tables of 1s and 2s.

The picture in the book is to help children visualize what multiplication is. You may use other pictures and real objects, too. Make up problems about them and have the children, also, try to make up problems. Most of the work should be done mentally and orally. If you use writing, have the children write down only the answers. For the reasons behind this, you may wish to reread the sections in this manual about teaching. Especially be familiar with the three stages of arithmetic growth.

Week 5

Lesson XLI

Objectives: to master the multiplication table of 3s.

Refer to the weekly schedule and follow it as closely as you can this week.

Weeks 6 to 7

Lessons XLII and XLIII

Objectives: to master the multiplication tables of 4s and 5s.

Teach these lessons using whatever procedures worked well for you during week 5. Remember to stress the mental and oral practice. If you do this regularly, you will obtain the best kind of testing and diagnosis possible with young children. Do your children do this multiplication work now with ease and understanding? Or do they still seem to be struggling with it?

At this time make a decision about which course to follow below. The basic course is for children who need to move slower in order to attain good understanding. It allows more time on the multiplication and division facts up to 5s. The higher facts can be learned next year. The advanced course is for children who understand and memorize the multiplication facts well. It continues the multiplication unit through the table of 10s.

BASIC COURSE

UNIT: DIVISION

Weeks 8 to 12. Use division Lessons LII to LVI, taking one lesson per week. Concentrate on the lower numbers with dividends of 20 or less.

ADVANCED COURSE

UNIT MULTIPLICATION

Weeks 8 to 15. For children who understand and memorize the multiplication facts well, continue the lessons in order through the rest of the multiplication unit. Take one lesson per week.

Basic Course
(continued)

UNIT: SIGNS

Weeks 13 to 17. Review addition and subtraction Lessons X to XXXVII, this time teaching *plus*, *minus*, and *equals* signs. Let the children write problems in sentence form like this.

$$4 + 2 = 6$$

Weeks 18 to 23. Repeat multiplication Lessons XXXVIII to XLIII, the tables from 1s to 5s. Teach the *times* and *equals* signs and let the children begin writing the problems in sentence form like this.

$$3 \times 4 = 12$$

UNIT: REVIEW

Weeks 24 to 36. Begin the signs unit on page 69. This reviews addition and subtraction. Move on through the reviews as far as your children are able to go this year.

Advanced Course
(continued)

UNIT: DIVISION

Weeks 16 to 29. With children who are mastering the work as you go, continue using one lesson per week. Follow the typical weekly schedule with whatever adjustments seem to fit your particular pupils. Remember to spend time each week on measurement activities and other life arithmetic situations.

UNIT: SIGNS AND REVIEW

Weeks 30 to 36. Teach the signs as on page 69, and for the rest of the school year let the children practice writing some of their problems in this sentence form.

Continue through the reviews and the tables, going as far as your children are able to this school year. Addition which moves into the next decade, as from the 30s to the 40s, should be practiced mentally or with an abacus or counters. On the tables, work particularly with U.S. money, avoirdupois weight, dry and liquid measures, long measure, time, and dozen.

PROGRESS RECORD
Grade 2

	count,	add	subtract	multiply		divide		works mostly at		
Names	read, write to 100	sums to 20	sums to 20	to 5s	to 10s	below 20	above 20	manipu-lative stage	mental-image stage	Comments

Testing Second Graders

In Ray's arithmetic, you should be leading the children almost daily in oral work. So you have a good assessment of their achievement as you go along, and you may not feel much need for tests. But from time to time you may want to test, anyway, to check up on each pupil's progress. The tests can diagnose for you by showing in what areas a pupil needs additional practice or reteaching.

Tests for each unit in second grade are given on the following pages of this manual. For best results, give the tests individually and orally, except for the test on signs. If this is not possible, give the tests in small groups so you can read the problems and see that each child writes his answers in the correct spaces. The answers may be written on blank papers. Or if you wish to make copies of the tests, you have the publisher's permission to do so while you are teaching from Ray's arithmetics.

Test Schedule

Unit 1: Review. Use any of the first grade tests.

With these tests you can diagnose how to spend time during the first three weeks. Give your second graders a good review and the rest of the year's work will go more smoothly.

Unit 2: Multiplication. Use the tests, "Multiplication Facts" and "Solving Multiplication Problems."

Use the facts test as often as you wish, for practice and for finding which facts a child should study. Each child can make his private set of flashcards for facts that he misses. Use the problems test at the end of the unit, and give it orally.

In your progress records, mark the level where each child works. For instance, mark whether he knows his multiplication tables up to 5s or 7s or 10s or whatever. And mark whether he can do problems in his head or if he needs manipulative objects.

Unit 3: Division. Use the tests, "Division Facts" and "Solving Division Problems."

Use the facts test as often as you wish, for practice and for checking up on progress. Use the problems test at the end of the unit. Give it orally.

Unit 4: Signs. Use the "Signs" test.

With this test, the children should have copies of the test sheet so they can read the signs for themselves and write their answers. It is the first test in Ray's arithmetics that need not be given orally.

Multiplication Facts

Name ______________________________ Date _______________

To 5s:

1 times 4 is ___________
1 times 1 is ___________
1 times 3 is ___________
1 times 2 is ___________
1 times 5 is ___________
5 times 5 are ___________
2 times 2 are ___________
2 times 5 are ___________
2 times 3 are ___________
3 times 5 are ___________
4 times 4 are ___________
3 times 3 are ___________
2 times 4 are ___________
4 times 5 are ___________
4 times 3 are ___________

Score ___________

To 10s:

7 times 2 are ___________
6 times 4 are ___________
9 times 2 are ___________
2 times 6 are ___________
8 times 6 are ___________
3 times 7 are ___________
1 times 6 are ___________
6 times 3 are ___________
7 times 6 are ___________
9 times 7 are ___________
3 times 8 are ___________
7 times 7 are ___________
10 times 4 are ___________
6 times 5 are ___________
10 times 3 are ___________
5 times 7 are ___________
6 times 6 are ___________
10 times 5 are ___________
9 times 5 are ___________
7 times 8 are ___________
6 times 9 are ___________
8 times 8 are ___________
9 times 1 are ___________
4 times 9 are ___________
7 times 4 are ___________
2 times 8 are ___________
8 times 5 are ___________
9 times 9 are ___________
8 times 4 are ___________
3 times 9 are ___________
10 times 6 are ___________
10 times 8 are ___________
10 times 2 are ___________
10 times 7 are ___________
10 times 9 are ___________
9 times 8 are ___________

Score ___________

Solving Multiplication Problems

Name ______________________________ Date ______________

Lower Numbers

1. If a car travels 1 mile in a minute, how far will it travel in 4 minutes? __________

2. If a piece of candy costs 5 cents, how much do 2 pieces cost? __________

3. If each pear is worth 5 apples, how much are 3 pears worth? __________

4. Thomas has 3 pigeons and James has 4 times as many. How many pigeons has James? __________

5. Lucy has 5 hens and each hen has 5 baby chicks. How many baby chicks are there? __________

Score ____________

Higher Numbers

6. If there are 6 panes of glass in one window, how many panes are there in 3 windows? __________

7. Edward has 2 pockets and 7 marbles in each. How many marbles does he have? __________

8. Seven children are in the class and each child needs 8 sheets of paper. How many sheets are needed all together?

____

9. In one dime there are 10 cents. How many cents are in 9 dimes? __________

10. If one toy truck costs 10 dollars, how much do 3 toy trucks cost? __________

Score ____________

Division Facts

Name ______________________________ Date _______________

Dividends below 20

2 in 2 ________	3 in 9 ________
3 in 15 ________	7 in 14 ________
2 in 10 ________	4 in 16 ________
4 in 12 ________	5 in 15 ________
2 in 4 ________	3 in 3 ________
4 in 8 ________	6 in 18 ________
2 in 18 ________	4 in 20 ________
6 in 6 ________	2 in 16 ________
5 in 10 ________	2 in 14 ________
3 in 18 ________	5 in 20 ________
7 in 7 ________	2 in 8 ________
9 in 18 ________	5 in 5 ________
8 in 16 ________	2 in 12 ________
2 in 6 ________	8 in 8 ________
6 in 12 ________	2 in 20 ________
3 in 6 ________	3 in 12 ________

Score ____________

Dividends above 20

4 in 24	________	9 in 81	________
7 in 63	________	3 in 27	________
9 in 36	________	6 in 30	________
5 in 25	________	8 in 56	________
4 in 36	________	7 in 28	________
9 in 90	________	5 in 40	________
6 in 24	________	4 in 28	________
3 in 30	________	5 in 35	________
4 in 32	________	7 in 70	________
5 in 30	________	9 in 45	________
6 in 48	________	7 in 56	________
5 in 45	________	6 in 42	________
7 in 49	________	8 in 72	________
8 in 64	________	7 in 21	________
4 in 40	________	9 in 72	________
9 in 63	________	3 in 24	________
6 in 54	________	8 in 40	________
7 in 35	________	9 in 54	________
9 in 27	________	8 in 80	________
8 in 32	________	3 in 21	________
6 in 36	________	8 in 24	________
8 in 48	________	7 in 42	________

Score ________

Solving Division Problems

Name ______________________________ Date _______________

Lower Numbers

1. Ten jacks are in a set. How many twos can a player pick up? __________

2. Three feet make a yard. How many yards are in 15 feet? __________

3. Four quarts make a gallon. How many gallons are 12 quarts? __________

4. If 16 apples are divided among 4 boys, how many apples will each boy have? __________

5. If 25 cents are divided among 5 girls, how much will each girl have? __________

Score ____________

Higher Numbers

6. There are 24 fruit trees in 6 rows. How many trees are in each row? __________

7. If 7 boys share 35 dollars equally, how much will each boy have? __________

8. If one doll costs 8 dollars, how many dolls can be bought for 48 dollars? __________

9. If you place 81 blocks in 9 rows, how many blocks will be in each row? __________

10. How many ten-dollar shirts can be bought for 70 dollars? __________

Score ____________

Signs

Name ______________________ Date ____________

Numbers to 20

$8 + 2 =$ ________	$8 \times 2 =$ ________
$6 \times 3 =$ ________	$6 - 3 =$ ________
$12 + 2 =$ ________	$12 \times 2 =$ ________
$15 - 5 =$ ________	$19 - 8 =$ ________
$8 + 2 =$ ________	$12 - 2 =$ ________
$6 + 3 =$ ________	$3 \times 4 =$ ________
$2 \times 4 + 3 =$ ________	$10 + 6 - 3 =$ ________
$8 + 8 - 6 =$ ________	

Score ________

Numbers above 20

$22 + 2 =$ ________	$64 + 2 =$ ________
$29 + 5 =$ ________	$66 + 3 =$ ________
$35 + 6 =$ ________	$40 + 10 =$ ________
$10 \times 3 =$ ________	$37 - 5 =$ ________
$29 + 2 =$ ________	$42 + 10 =$ ________
$25 - 5 =$ ________	$26 - 4 =$ ________
$6 + 10 - 3 =$ ________	$20 + 10 - 6 =$ ________
$5 \times 5 + 5 =$ ________	$6 \times 6 - 6 =$ ________
$27 + 10 - 5 =$ ________	$13 + 10 + 4 =$ ________

Score ________

Answer Key

MULTIPLICATION FACTS. To 5s, Column 1: 4, 1, 3, 2, 5, 25, 4, 10. **Column 2:** 6, 15, 16, 9, 8, 20, 12. **To 10s Column 1:** 14, 24, 18, 12, 48, 21, 6, 18, 42, 63, 24, 49, 40, 30, 30, 35, 36, 50. **Column 2:** 45, 56, 54, 64, 9, 36, 28, 16, 40, 81, 32, 21, 60, 80, 20, 70, 90, 72.

SOLVING MULTIPLICATION PROBLEMS. Lower Numbers: 1) 4. **2)** 10. **3)** 15. **4)** 12. **5)** 25. **Higher Numbers: 6)** 18. **7)** 14. **8)** 56. **9)** 90. **10)** 30 dollars.

DIVISION FACTS. Dividends below 20, Column 1: 1, 2, 2, 3, 2, 2. **Column 2:** 3, 2, 4, 3, 1, 3, 5, 8, 7, 4, 4, 1, 6, 1, 10, 4. **Dividends above 20, Column 1:** 6, 9, 4, 5, 9, 10, 4, 10, 8, 6, 8, 9, 7, 8, 10, 7, 9, 5, 3, 4, 6, 6. **Column 2:** 9, 9, 5, 7, 4, 8, 7, 7, 10, 5, 8, 7, 9, 3, 8, 8, 5, 6, 10, 7, 3, 6.

SOLVING DIVISION PROBLEMS. Lower Numbers: 1) 5. **2)** 3. **3)** 3. **4)** 4. **5)** 5. **Higher Numbers: 6)** 4. **7)** 5. **8)** 6. **9)** 9. **10)** 7.

SIGNS. Numbers to 20, Column 1: 10, 18, 14, 10, 10, 9, 11, 10. **Column 2:** 16, 3, 24, 11, 10, 12, 13. **Numbers above 20, Column 1:** 24, 34, 41, 30, 31, 20, 13, 30, 32. **Column 2:** 66, 69, 50, 32, 52, 22, 24, 30, 27.

Teaching Third Grade

Using *Ray's New Intellectual Arithmetic*

page

Teaching Third Grade

The main objectives of the third grade curriculum are to gain ability to work with higher numbers in addition and subtraction; to work up to the 12s in multiplication and division; to learn ways of writing out problems for all basic operations; and to read, write, and mentally solve problems with simple fractions. This covers Lessons I through XXVI in *Ray's Intellectual Arithmetic.*

Children who have been trained in the mental work of Ray's arithmetics in grades 1 and 2 should now easily make the transition to written forms. But don't omit the mental practice. During this year keep a good balance between oral work and written work.

Also keep a good balance between understanding and memorizing. Children should understand concepts and principles rather than just memorizing them. And they should understand the problems they do. But also give plenty of practice and drill on the basic facts so that children memorize them to the point of overlearning.

In the schedules this year, you will often be spending several days on one lesson of the textbook. By the last review of such a lesson, both you and the children should see how much easier it is for the pupils. Commend them for sticking with something until they learn it well. This method will build in them good study habits. It also will help them see that they are progressing and learning. This is something that many young children do not recognize without help from their teachers.

The parent or teacher who works daily with a child is the best judge of when to move along more rapidly or when to slow down and give more practice. So the guides and schedules given below need not be followed rigidly. Use them as general guidelines to help you achieve the most you can during this year with your pupils.

A Typical Day

Schedule arithmetic in the morning when the children are mentally alert. Spend the first part of the period with the main work for the day. This will usually be on the content of the week's lesson, sometimes from the textbook and sometimes with flashcards or other teaching materials which give practice on the lesson. End that part of the work before the children tire from concentrating. Then for the rest of the period, assign a project such as those suggested in the project section of this manual. Or play some number games together, as suggested in the game section of the manual. Or let the children play among themselves. These activities are a valuable part of your arithmetic teaching, as they help your children develop a good understanding of numbers.

A Typical Week's Schedule

Monday. Use the textbook lesson orally to begin teaching the new content and to diagnose what the children should work on the most during the week. On a long lesson, you may complete only a part of it.

Tuesday. Use manipulative objects, flashcards, games, or any other means to practice on the content of the new lesson.

Wednesday. Try the textbook lesson again. You may have the children write out the problems this time. Notice improvement in their understanding and diagnose any trouble spots.

Thursday. Repeat practice as on Tuesday. On long lessons, you may continue work in the textbook from wherever you left off. If the lesson is well mastered, you may use this day for extra projects, as in the project section of this manual.

Friday. Review the textbook lesson. Make notes for yourself about anything you want to provide additional practice on next week.

Planning Guide for Grade 3

(Ray's Intellectual Arithmetic)

UNIT: ADDITION

Weeks 1 to 7

Lessons I to V

Basic Objectives: to master addition facts through the 12s and to understand and use these facts in solving problems.

Advanced Objective: to compute addition problems with carrying.

Begin the year with a check-up on reading and writing numbers to 100 or to 1000, depending on the ability of the pupils. Work through the addition lessons, sometimes mentally, sometimes with manipulative objects, and sometimes using written forms. Notice that some of the problems now involve adding across the decades. That is, the children may be asked to add to a number in the 50s and get a sum in the 60s. Use the counting game Zip, and in other ways practice this skill often. When writing problems, teach both the sentence form and the vertical form, as shown below.

$$8 + 5 + 3 = 16 \qquad\qquad \begin{array}{r} 8 \\ 5 \\ +\ 3 \\ \hline 16 \end{array}$$

Spend time each week on the tables in Lesson III. These should be overlearned. Use practice drills, games, and tests to accomplish this. For advanced pupils, spend some time on sections 18 and 19, beginning on page 24 of *Ray's Practical Arithmetic*. This will acquaint them with the procedure of carrying.

UNIT: SUBTRACTION

Weeks 8 to 14

Lessons VI to XI

Basic Objectives: to master subtraction facts through the 12s and to understand and use these facts in solving problems.

Advanced Objective: to compute subtraction problems with borrowing.

Work through the subtraction lessons in a variety of ways, as in the addition unit. Use manipulative objects and written forms, as well as oral practice. Teach both the sentence form and the vertical form, as shown below.

$17 - 5 = 12$

$$\begin{array}{r} 17 \\ -\ 5 \\ \hline 12 \end{array}$$

Every week spend time on the tables in Lesson VII so that the pupils overlearn these facts. Teach that some subtraction answers are *differences* and some are *remainders*. In other words, some subtraction problems involve finding differences in heights, ages, or other amounts; and some involve finding a remainder after an amount is taken away. If your children understand those two kinds of problems it will make subtraction easier for them. Advanced pupils may spend some time on sections 25 and 26 beginning on page 33 of *Ray's Practical Arithmetic*. This will introduce them to the concept of borrowing.

UNIT: MULTIPLICATION

Weeks 15 to 23

Lessons XII to XV

Objectives: to master the multiplication tables

through the 12s and to use these facts in solving problems.

Spend time every week on the tables in Lesson XIII so that pupils overlearn them. Use a variety of drills, games, and speed tests to help the children practice. Problems in the other lessons may be done with manipulative objects one day, orally another day and repeated in written form a third day. Teach two forms for writing the problems.

$3 \times 11 = 33$

$$\begin{array}{r} 11 \\ \times\ 3 \\ \hline 33 \end{array}$$

Continue to work on measurement skills. Also use a variety of ideas from the games and projects sections of this manual to provide plenty of life experience in using numbers.

UNIT: DIVISION

Weeks 23 to 30

Lessons XVI to XIX

Objectives: to master the division tables through the 12s and to use these facts in computing and solving problems.

Use a variety of approaches as in the multiplication unit. Spend time every week drilling on the tables in Lesson XVII, so pupils overlearn these facts. Teach three forms for writing division problems.

1) sentence form: $120 \div 10 = 12$

2) box form with quotient written either above or to the right:

$$10\overline{)\,120\,}(12$$

3) fraction, or ratio, form:

$$\frac{120}{10}$$

UNIT: FRACTIONS

Weeks 31 to 36

Lessons XX to XXVI

Objectives: to be able to read and write simple fractions and to understand their meaning well enough to solve problems mentally and orally.

Begin the lessons in this fractions unit and work as far as the children are able to. Help them understand two uses of fractions: 1) fractional parts of a unit and 2) fractional parts of a group. They can split an apple and get halves, or they can divide the class into halves for teams. You need not complete the fractions unit this year, but get your children started. Next year they may review and relearn as needed.

PROGRESS RECORD

Grade 3

	read, write	add		subtract		multiply		divide		fractions		
Names	to 100	to 12s	use + and = signs	to 12s	use signs	to 12s	use signs	to 12s	use signs	read and write	solve problems	Comments

Testing Third Graders

During third grade, children should become proficient in the four basic operations of arithmetic: addition, subtraction, multiplication, and division. So in your testing program you may want to use the tests on basic facts given in the first and second grade sections of this manual. These can be used as often as you wish, for diagnosis and for speed practice.

Tests for each unit are given on the following pages of this manual. Use them at the end of the units, to check-up on learning. When necessary, they may also be used for placement purposes. For instance, if you have new pupils joining your class or if your children are starting on Ray's system of arithmetic for the first time, the tests can help you decide what units they should begin studying. You have the publisher's permission to make copies of the tests if you are using Ray's arithmetics.

Test Schedule

Unit 1: Addition. Before beginning the unit, you may give the first grade tests on addition facts and solving addition problems. This will show whether the children have ability in adding smaller numbers, which they need before doing this new unit on adding higher numbers. If these tests show the need for it, do some remedial, catch-up work.

After the unit, use the following addition test to see how well your children know the new material. You will notice that only a few of the problems use the plus and equals signs. That is so any children who have just begun using signs will be able to show how well they can add.

Unit 2: Subtraction. Before the unit, you may wish to give the first grade subtraction tests to determine if your children are ready to subtract with higher numbers. Provide extra review and practice if necessary. After the unit, give the third grade subtraction test to see how well they have learned the material in this unit.

Unit 3: Multiplication. Before the unit, you may check up on your children's knowledge and skills with smaller numbers by giving the second grade tests on multiplication facts and solving multiplication problems. This will show whether your group needs to practice a lot on the smaller numbers or whether they are ready to go on to things like multiplying by 11 and 12. After the unit, give the third grade multiplication test, which follows in this manual. This will show how much your children have learned during the unit.

Unit 4: Division. Before the unit you may give the second grade tests on division to see how well the children remember those facts and skills. This will show whether your children need extra work on the smaller numbers as you study this unit or whether they are ready to do problems with higher numbers. After the unit, give the third grade division test which follows in these pages. It checks up on the tables of 10, 11, and 12; and problem solving with division.

Unit 5: Fractions. This test will help you determine if your children can read and write simple fractions and if they understand fractional parts of a single unit and fractional parts of a group. With these skills you can be assured they have a good foundation for further work in fractions. Notice that the first part of the test must be read orally.

Notice, also, that according to the planning guide you do not have to finish the unit on fractions this year. But you can do a few of the lessons and work with real objects. Then give the test to see how well your children understand fractions.

Addition

Name ______________________ Date ____________

Add:

8 and 3 _____	28 and 3 _____	48 and 3 _____
9 and 7 _____	49 and 7 _____	69 and 7 _____
14 and 4 _____	24 and 4 _____	84 and 4 _____
6 and 10 _____	16 and 10 _____	76 and 10 _____
5 and 11 _____	25 and 11 _____	55 and 11 _____

19 and 8 and 6 _______	8 and 5 and 3 _______
7 and 10 and 9 _______	11 and 7 and 5 _______
10 and 8 and 6 _______	27 and 4 and 3 _______

Count by 2s to 20 ______________________________

Count by 5s to 20 ______________________________

Count by 8s to 80 ______________________________

____

25 + 6 = ____	38 + 3 = ____	49 + 5 = _____
7 + 12 = ____	27 + 12 = ____	10 + 8 = _____
32 + 5 = ____	61 + 9 = ____	28 + 5 = _____

Score __________

Subtraction

Name ____________________________ Date ______________

6 − 3 = _____	9 − 4 = ____	8 − 7 = ____
17 − 7 = _____	15 − 4 = ____	18 − 9 = ____
21 − 9 = _____	19 − 10 = ____	10 − 9 = ____
14 − 5 = _____	25 − 14 = ____	12 − 6 = ____
27 − 8 = _____	22 − 10 = ____	100 − 5 = ____

9 − 5 + 7 − 2 = _____ 8 − 5 + 4 − 2 = _____

6 − 2 + 4 − 3 + 2 = ______

7 − 4 + 5 − 4 + 3 − 2 = _____

20. A man owed 60 dollars. He paid 20 dollars and then another 30 dollars. Later he borrowed 5 dollars. How much does he still owe? ________

21. An orchard has 5 apple trees and 11 peach trees in one row, and 8 apple trees and 8 peach trees in another row. How many more peach trees than apple trees are there? ________

22. Thomas had five dollars, but lost 2 of them. How much money did he have left? ________

23. Sheila is 9 years old, and her sister Grace is 17 years old. How much older is Grace? ________

24. Tom is 8 years old and his father is 32 years old. In how many years will Tom be as old as his father is now? ________

25. Harry had 18 marbles and gave away 10. How many does he have left? ________

Score __________

Multiplication

Name ______________________ Date ____________

11 × 2 = ____	12 × 3 = ____	5 × 7 = ____
12 × 8 = ____	10 × 10 = ____	10 × 3 = ____
11 × 4 = ____	12 × 12 = ____	11 × 9 = ____
10 × 9 = ____	10 × 12 = ____	11 × 3 = ____
11 × 7 = ____	11 × 11 = ____	12 × 9 = ____
12 × 5 = ____	11 × 12 = ____	12 × 7 = ____
2 × 12 = ____	4 × 12 = ____	6 × 12 = ____
7 × 7 = ____	10 × 11 = ____	5 × 11 = ____

25. There are 3 columns above and 8 problems in each column. How many problems are there? ________

26. On a checkerboard are 8 rows of squares, and 8 squares in each row. How many squares are on the board? ________

27. If a man runs a mile in 6 minutes, how long will it take him to run 3 miles? ________

28. If 4 men can do a piece of work in 6 days, how many days will it take 1 man to do it? ________

29. What is 3 times the sum of 5 and 7? ________

30. Three dozen eggs will serve how many campers 1 egg each? ________

Score ________

Division

Name ______________________________ Date ______________

33 ÷ 11 = ________	20 ÷ 2 = ________
80 ÷ 8 = ________	100 ÷ 10 = ________
24 ÷ 2 = ________	132 ÷ 11 = ________
96 ÷ 8 = ________	108 ÷ 9 = ________
132 ÷ 12 = ________	88 ÷ 11 = ________
110 ÷ 10 = ________	30 ÷ 3 = ________
90 ÷ 9 = ________	120 ÷ 10 = ________
44 ÷ 4 = ________	66 ÷ 11 = ________
110 ÷ 11 = ________	50 ÷ 5 = ________
99 ÷ 9 = ________	70 ÷ 10 = ________
60 ÷ 6 = ________	55 ÷ 5 = ________
121 ÷ 11 = ________	72 ÷ 6 = ________
36 ÷ 12 = ________	77 ÷ 11 = ________
22 ÷ 2 = ________	48 ÷ 4 = ________
40 ÷ 4 = ________	120 ÷ 12 = ________
84 ÷ 7 = ________	60 ÷ 5 = ________
144 ÷ 12 = ________	

34. Jake has 40 dollars and he uses 5 dollars a week. How many weeks will the money last him? ________

35. William does 10 push-ups while Sam does 7. How many push-ups can Sam do while William does 60? ________

36. Cindy has saved 84 dollars. That is 12 times as much as her little sister has saved. How much has her sister saved? ________

37. I am thinking of a number, which divided by 3 is the same as 2 times 6. What is my number? ________

38. There are 121 trees in an orchard, and 11 rows of trees. How many trees are in each row? ________

39. Bob counted 144 eggs in a carton. How many dozen is that? ________

40. How many dozen are in 72? ________

Score __________

Fractions

Read these fractions.

1/2   1/3   1/4   3/4   2/3   7/10   3/10   75/100

Write these fractions.

two-thirds ________   fifteen-one hundredths ________

four-fifths ________   five-twelfths ________

five-tenths ________   six-sevenths ________

15. How many halves are in an apple? ________

16. How many halves are in 2 apples? ________

17. How many fourths are in an apple? ________

18. How many fourths are in 3 apples? ________

19. A class had 24 members. Half of them were sick with flu. How many were sick? ________

20. If you have a dozen candy bars to share among three children, how many bars will each of you get? ________

21. If you share a dozen candy bars with one friend, how many bars will each of you get? ________

22. If you cut a cake for 16 people, what fraction is each part called? ________

23. If you cut the cake for 20 people, what fraction is each part called? ________

24. If you miss 5 of these 24 problems, what fractional part will you miss? ________

Score ________

Answer Key

ADDITION. How many, Row 1: 11, 31, 51. **Row 2:** 16, 56, 76. **Row 3:** 18, 28, 88. **Row 4:** 16, 26, 86. **Row 5:** 16, 36, 66. **Row 6:** 33, 16. **Row 7:** 26, 23. **Row 8:** 24, 34. **Count by 2s:** 2, 4, 6, 8, 10, 12, 14, 16, 18, 20. **Count by 5s:** 5, 10, 15, 20. **Count by 8s to 80:** 8, 16, 24, 32, 40, 48, 56, 72, 80. **Add, Row 1:** 31, 41, 54. **Row 2:** 19, 39, 18. **Row 3:** 37, 70, 33.

SUBTRACTION. Column 1: 3, 10, 12, 9, 19. **Column 2:** 5, 11, 9, 11, 12. **Column 3:** 1, 9, 1, 6, 95. **Add and Substract:** 9, 5, 7, 5. **Story Problems: 20)** 15 dollars. **21)** 6. **22)** 3 dollars. **23)** 8. **24)** 24. **26)** 8.

MULTIPLICATION. Tables: Column 1: 22, 96, 44, 90, 77, 60, 24, 49. **Column 2:** 36, 100, 144, 120, 121, 132, 48, 110. **Column 3:** 35, 30, 99, 33, 108, 84, 72, 55. **Problems: 25)** 24. **26)** 64. **27)** 18. **28)** 24. **29)** 36. **30)** 36.

DIVISION. Tables: Column 1: 3, 10, 12, 12, 11, 11, 10, 11, 10, 11, 10, 11, 3, 1 1, 10, 12, 12. **Column 2:** 10, 10, 12, 12, 8, 10, 12, 6, 10, 7, 11, 12, 7, 12, 10, 12. **Problems: 34)** 8. **35)** 42. **36)** 7 dollars. **37)** 36. **38)** 11. **39)** 12. **40)** 6.

FRACTIONS. Read: one half, one third, one fourth, three fourths, two thirds, seven tenths, three tenths, seventy-five (one) hundredths. **Write, Column 1:** 2/3, 4/5, 5/10. **Column 2:** 15/100, 5/12, 6/7. **Problems: 15)** 2. **16)** 4. **17)** 4. **18)** 12. **19)** 12. **20)** 4. **21) 6. 22)** 1/16. **23)** 1/20. **24)** 5/24.

Teaching Fourth Grade

Using *Ray's New Intellectual Arithmetic*

page

Teaching Fourth Grade

The main objectives of the fourth grade curriculum are to learn to work with fractions, ratios, and percentages. Children should also continue to grow in their skills in addition, subtraction, multiplication, and division, and in their knowledge of measurements. This work covers Lessons XX to LXXX in *Ray's Intellectual Arithmetic*.

During this year, keep a good balance between oral and written work. Children will profit from both methods. Also use manipulative objects, particularly when introducing new work in fractions. You may wish to reread information about the three stages of arithmetic growth which are described in the section on tips for teaching Ray's arithmetics. Children need time in the manipulative stage whenever they are learning a new topic.

Teaching fractions and ratios and percentages this year will keep you busy. There is a lot for the children to learn. So it is best to schedule arithmetic in the mornings when the children are more alert mentally.

The planning guide below includes more than one lesson per week, so it may appear to move along faster than in the primary years. Notice the flexibility built into the plan, especially on the topics of ratio and percentage. Make use of the flexibility to fit the plan to your particular children.

A Typical Day

Spend the first part of the period teaching the new lesson content or reteaching something difficult from the previous day. In fourth grade, you may be able to spend most or even all of the period on such study. But if your children tire from the concentration, you may plan to use part of the period on projects or games as suggested elsewhere in this manual. In this more relaxed way your children will still improve their arithmetic skills.

A Typical Week's Schedule

Monday. Use the first textbook lesson for the week, working orally most of the time. Sometimes individual children may demonstrate on the chalkboard how they got the answers either by drawings of objects which they have visualized in their heads or by writing out the problems. Remember to use manipulative objects anytime it will help your children understand better.

Tuesday. Continue with the lesson begun on Monday. Also, you may repeat some of Monday's problems, having the children write them out this time. This makes a good check on their learning.

Wednesday. If the first lesson for the week is learned well, begin the next lesson. If not, spend a third day on the lesson begun on Monday.

Thursday. This will be similar to Monday's schedule or similar to Tuesday's, depending on whether you started a new lesson on Wednesday.

Friday. Review all the work for the week. Finish lessons as needed. Spend part of the period on a project or game.

Planning Guide for Grade 4

(Ray's Intellectual Arithmetic)

UNIT: FRACTIONS

Weeks 1 to 12

Lessons XX to XLIV

Objectives: to read, write, and compute fractions and mixed numbers; to understand the terms *denominator*, *numerator*, *proper fraction*, *improper fraction*, *lowest terms*, *common denominator*, and *least common denominator*.

Fractions are one of the major topics to master this year, so spend about twelve weeks on this unit. For a basic course, you can cover one lesson per week, using Lessons XX through XXX and Lesson XLIV. For an advanced course you can cover about two lessons per week and do most or all lessons in the unit. If you follow the textbook carefully, you will notice how the orderly arrangement of each topic leads the children to discover rules by which to work the problems. In each lesson, use as many problems as will fit your children's time and need for drill.

In both the basic and advanced course, be sure all children study Lesson XLIV. The concepts and tables in this lesson are extremely useful in everyday arithmetic.

Children will have an easier time with fractions if you first make sure they know the multiplication and division facts and if they have facility in counting activities, such as on page 16. So review these matters. Use flashcard reviews and the game Zip.

UNIT: REVIEW

Weeks 13 to 14

Lessons XLV and XLVI

Objectives: to review the four basic operations and fractions by using them in solving problems.

Moving at the rate of about twelve problems per day, you can cover these lessons in two weeks. But it is not necessary to work all twelve problems each day. Assign as many problems as will fit the allotted time. Diagnose trouble spots any of your children may have, and do any reteaching that may be indicated.

UNIT: TABLES

Weeks 15 to 16

Lessons XLVII to LII

Objectives: to master tables of money and measurements and be able to use the facts in solving problems.

Choose the tables you feel are most useful for your children and use a couple of days studying each of them. When possible, bring to class any items the children are not familiar with. For example, local coin collectors may show some eagles. A mill is "money of account," which means it is used, for instance, in computing property taxes, even though there is no coin.

UNIT: REVIEW

Weeks 17 to 19

Lessons LIII to LIX

Objectives: to review measures, fractions, and the four basic operations by using them in solving problems.

Moving at about eleven problems per day you can cover these lessons in three weeks. But it is not necessary to assign all eleven problems; use only as many as time allows. Spend time talking about the problems. Have the children explain how they arrived at their solutions. This way you will be testing and diagnosing every day. And you can reteach whenever you see a need to.

UNIT: RATIO

Weeks 20 to 23

Lessons LX to LXIII

Objectives: to understand the concept of ratio and be able to work problems involving ratios.

Teach what ratio is, using Lesson LX. For a basic course, each day work some of the easier problems in the unit, and repeat some of the same problems on subsequent days. Encourage the children to explain the problems by manipulating objects. Emphasis is on understanding ratios and fractional parts and not on writing the numbers in a particular way. Let the children explain and defend their own means of arriving at the answers.

For an advanced course, spend time together solving some of the difficult, puzzle-like problems. This is good training in thinking skills, as well as in arithmetic. Problems like these are sometimes used in tests of academic or mental ability. There are more such problems in Lesson LXXX to use with children who enjoy them.

UNIT: PERCENTAGE

Weeks 24 to 36

Lessons LXIV to LXXX

Basic Objectives: to be able to read, write, and compute percents of numbers up to 100.

Advanced Objectives: to be able to read and write percents, and to compute with them in problems involving discounts, commissions, premiums, and interest.

For the basic course, work at least through Lesson LXVIII. For the advanced course, work more lessons, through Lesson LXXII if possible. All children will profit by a review of Lesson XLIV in the fractions unit. A good knowledge of these fraction and percent equivalents can make problem solving easier and faster. The last lesson in the book is a review of all topics studied so far. So try to leave some time to spend on that lesson.

PROGRESS RECORD
Grade 4

	fractions and mixed numbers					tables of measures						ratio		percent	
Names	read and write	reduce	common denom- inators	add	multiply	money	dry	liquid	weight	length	time	read and write	solve problems	compute	solve problems

Testing Fourth Graders

Fourth graders' major job for the year is mastering fractions. They also should master the measurement tables and learn to solve problems with ratios and percent. Tests on these four topics are given on the following pages. You may copy them for use with your pupils.

There are two review units this year to help keep all skills sharp. No specific tests are included here for the review units, but you may wish to use tests from previous grades to check up on skills and knowledge learned previously.

Test Schedule

Unit 1: Fractions. Before teaching this unit, you can give the third grade fractions test to see if your children have an understanding of simple fractions. If this test shows they need review, use many manipulative materials to help them gain understanding. During the long unit, you may want to make up short tests from time to time to see if the children have mastered particular concepts. Near the end of the unit, give the fractions test which follows in this manual. From it you will easily see which kinds of problems the children may need more help with.

Unit 2: General Review. This unit has much review of fractions as well as the basic operations with whole numbers. If your children did not do as well as you would like on the fractions test, try giving it again after this review to see their progress in understanding fractions.

Also during the study of this unit, you may wish to use some of the third grade tests on the four basic operations with whole numbers. These will help you diagnose where any reteaching may be needed.

Unit 3: Tables. Use this unit test to help motivate the children to memorize any of the tables they do not know. From the test results, both they and you will easily see which tables they need to study more.

Unit 4: General Review. This unit reviews measurements from Unit 3 as well as all basic operations with whole numbers and with fractions. After the unit, you may give again the Unit 1 or Unit 3 test to see how much better the children score this time.

Unit 5: Ratio. After the unit, use the test on ratio which follows on these pages. It will show whether your children have a basic understanding of the concept of ratio.

Unit 6: Percentage. With this unit you can use the percentage test which follows in these pages. After the test, help the children figure out what per cent their scores are. Also, you can do some reteaching on the basis of the problems missed. For instance, if several of your children missed problems 12 to 17 on changing fractions to per cents, have them study again Lesson LXVI on that topic. And you could have them memorize again the fractional parts of 100 as found on page 79 of the textbook.

Fractions

Name ______________________________ Date ______________

$\frac{1}{10}$ $\frac{2}{3}$ $\frac{6}{5}$ $\frac{3}{2}$ $\frac{8}{10}$ $\frac{10}{8}$

Write the proper fractions here. ____________________________

Write the improper fractions here. ___________________________

Reduce each fraction to its lowest terms.

$\frac{2}{6}$ = __________ $\frac{5}{10}$ = __________ $\frac{18}{20}$ = ________

$\frac{6}{12}$ = __________ $\frac{3}{6}$ = __________ $\frac{8}{10}$ = __________

Change each pair to fractions with a common denominator.

$\frac{2}{3}$ and $\frac{3}{4}$ ________ ________ $\frac{1}{2}$ and $\frac{1}{4}$ ________ ________

$\frac{2}{3}$ and $\frac{2}{5}$ ________ ________ $\frac{2}{5}$ and $\frac{5}{6}$ ________ ________

Reduce each improper fraction to a mixed number.

$\frac{5}{4}$ = __________ $\frac{9}{4}$ = __________ $\frac{3}{2}$ = __________

$\frac{10}{3}$ = __________ $\frac{9}{8}$ = __________ $\frac{7}{5}$ = __________

One-half and $\frac{3}{4}$ = ________ One-fourth and $\frac{1}{3}$ = ____

Three-sixths less $\frac{2}{6}$ = ____ One-half less $\frac{1}{3}$ = ________

Six times $\frac{3}{4}$ = __________ Four times $\frac{7}{8}$ = __________

Five times $3\frac{1}{2}$ = ________ Three times $4\frac{1}{3}$ = ________

Two times $6\frac{1}{8}$ = ________ Four times $2\frac{1}{10}$ = ________

$\frac{1}{2}$ of $\frac{1}{5}$ = ________ $\frac{1}{3}$ of 6 = __________

$\frac{1}{6}$ of $\frac{1}{6}$ = ________ $\frac{2}{3}$ of 12 = ________

$\frac{1}{4}$ of 100 = ________ $\frac{1}{8}$ of 100 = ________

$\frac{1}{3}$ of 100 = ________ $\frac{1}{5}$ of 100 = ________

$\frac{1}{3}$ of $\frac{1}{2}$ = ________ $\frac{3}{4}$ of $\frac{7}{8}$ = ________

$\frac{3}{4}$ of 100 = ________ $\frac{1}{2}$ of 100 = ________

Score __________

Tables

Name ______________________________ Date ______________

1. There are ______________ cents in one dime.
2. There are ______________ dimes in one dollar.
3. There are ______________ pints in one quart.
4. There are ______________ pecks in one bushel.
5. There are ______________ quarts in one gallon.
6. There are ______________ ounces in one pound.
7. There are ______________ pounds in one ton.
8. There are ______________ inches in one foot.
9. There are ______________ feet in one yard.
10. There are ______________ yards in one rod.
11. There are ______________ rods in one mile.
12. There are ______________ hours in one day.
13. There are ______________ days in one common year.
14. There are ______________ days in one leap year.
15. There are ______________ years in one century.
16. There are ______________ days in one month.
17. Two feet are what part of a yard? ______________
18. Six inches are what part of a foot? ______________
19. Three months are what part of a year? ______________
20. Thirty seconds are what part of a minute? ______________

Score ____________

Ratio

Name ______________________________ Date _______________

1. What is the ratio of 2 to 8? _______________

2. What is the ratio of 5 to 10? _______________

3. What is the ratio of 12 to 36? _______________

4. What is the ratio of 10 to 100? _______________

5. What is the ratio of 5 to 4? _______________

6. What is the ratio of 18 to 9? _______________

7. A class of 35 children has a ratio of 2 boys to 3 girls. How many of each are in the class? __________

8. Divide 20 pieces of candy so that the winning team gets 4 pieces for every 1 the losing team gets. __________

9. Jerry and Tim bought a hog together, and Jerry put in twice as much money as Tim. How much should each boy get if they sell the hog for 90 dollars? __________ __________

10. Mr. Smith invested 4 dollars in a company for every 1 dollar Mr. Jones invested. If their stock gains 800 dollars, how much will each man's share be? __________ __________

Score _____________

Percentage

Name ______________________________ Date ______________

Six per cent of 100 is ________ Six per cent of 50 is ________

Ten per cent of 20 is________ of 30________ of 15________

What part is 75 per cent? ________ 12½ per cent? ________

What part is 5 per cent? ________ 60 per cent? ________

What part is 30 per cent? ________ 20 per cent? ________

How many per cent is ½? ________ ⅓ ________ ⅔ ________

How many per cent is ¼?________ $^4/_{10}$________ $^3/_{10}$________

18. There were 50 pupils enrolled in a school. One day 5 of them were absent. What per cent were absent? __________

19. When the flu was going around, 8 pupils were absent from the school. What per cent was that? __________

20. A man bought some merchandise for 60 dollars and sold it for 69 dollars. What per cent did he gain? __________

21. A librarian bought some books priced at 500 dollars, but she got them for 15 per cent off. What did she pay for the books? __________

22. A salesman sold 560 dollars worth of goods, and his commission is 5 per cent. How much money did he earn? __________

23. At 18 per cent annual interest, how much is the interest on each 100 dollars borrowed? __________

24. If you get 20 problems correct on a test of 25 problems, what per cent is your score? __________

25. If you get all the problems correct, what per cent is your score? __________

Score ____________

Answer Key

FRACTIONS. Proper fractions: $\frac{1}{10}$, $\frac{2}{3}$, $\frac{1}{2}$, $\frac{8}{10}$. **Improper fractions:** $\frac{6}{5}$, $\frac{10}{8}$. **Reduce:** $\frac{1}{3}$, $\frac{1}{2}$, $\frac{9}{10}$, $\frac{1}{2}$, $\frac{1}{2}$, $\frac{4}{5}$. **Common denominators:** $\frac{8}{12}$ and $\frac{9}{12}$; $\frac{2}{4}$ and $\frac{1}{4}$; $\frac{10}{15}$ and $\frac{6}{15}$; $\frac{12}{30}$ and $\frac{25}{30}$. **Mixed numbers:** $1\frac{1}{4}$; $2\frac{1}{4}$; $1\frac{1}{2}$; $3\frac{1}{3}$; $1\frac{1}{8}$; $1\frac{2}{5}$. **Problems, left to right:** $1\frac{1}{4}$; $\frac{7}{12}$; $\frac{1}{6}$; $\frac{1}{6}$; $4\frac{1}{2}$; $3\frac{1}{2}$; $17\frac{1}{2}$; 13; $12\frac{1}{4}$; $8\frac{2}{5}$; $\frac{1}{10}$; 2; $\frac{1}{36}$; 8; 25; $12\frac{1}{2}$; $33\frac{1}{3}$; 20; $\frac{1}{6}$; $\frac{21}{32}$; 75; 50.

TABLES. 1) 10. **2)** 10. **3)** 2. **4)** 4. **5)** 4. **6)** 16. **7)** 2000. **8)** 12. **9)** 3. **10)** $5\frac{1}{2}$. **11)** 320. **12)** 24. **13)** 365. **14)** 366. **15)** 100. **16)** 30. **17)** $\frac{2}{3}$. **18)** $\frac{1}{2}$. **19)** $\frac{1}{4}$. **20)** $\frac{1}{2}$.

RATIO. 1) 1:4. **2)** 1:2. **3)** 1:3. **4)** 1:10. **5)** $1\frac{1}{4}$. **6)** 2. **7)** 14 and 21. **8)** 16 and 4. **9)** 60 and 30 dollars. **10)** 640 and 160 dollars.

PERCENTAGE. Per cents: 6; 3; 2; 3; $1\frac{1}{2}$; $\frac{3}{4}$; $\frac{1}{8}$; $\frac{1}{20}$; $\frac{6}{10}$; $\frac{3}{10}$; $\frac{2}{10}$; 50; $33\frac{1}{3}$; $66\frac{2}{3}$; 25; 40; 30. **18)** 10. **19)** 16. **20)** 15. **21)** 425 dollars. **22)** 28 dollars. **23)** 18 dollars. **24)** 80. **25)** 100 per cent.

Teaching Fifth Grade

Using *Ray's New Practical Arithmetic*

page

Teaching Fifth Grade

The objectives for fifth grade include being able to do the four basic operations with whole numbers, fractions, and mixed numbers; also to gain facility in factoring and working with compound numbers in measures. This covers sections 1 to 130 in *Ray's Practical Arithmetic*.

Children who have studied Ray's arithmetics in their early years should have good problem solving skills by fifth grade. This year you can help them continue to develop these skills. This is perhaps more important than ever in our electronic age. As calculators and computers become a part of your children's lives, they do not diminish the need to understand arithmetic; they increase it. When the children can work problems entirely in their heads, let them. At other times they may write out their work and you can check on their procedures.

If you are a parent-teacher who happens to have been away from arithmetic for a long time, do not panic. As you work along through the lessons with your child, you will find your own skills being sharpened. Don't be afraid of studying a page more than once. Many times children need repeated exposures to a new topic in order to understand it well. As you help them through this process, their own study skills will grow. They will learn not to back off from something hard, but to stick with it until they learn it.

Use the following schedules and guides to help in your own planning. Follow the plan as closely as you can, but adjust wherever necessary for your children.

A Typical Day

One good group plan is to work on the lesson all together at first. Then use part of the period for individualizing. In that time each child or group of children can work on materials you have helped them choose. For instance, those who have not mastered basic number facts can practice or play games with flashcards of the facts. More advanced children can work on puzzles or projects for enrichment. Vary the assignments, and make sure the slower pupils have as much fun as the advanced pupils.

It is not necessary to do the individualizing every day, but doing it once or twice a week helps keep all children interested in arithmetic and growing at their own speeds.

A Typical Week's Schedule

Monday. Study a new section in the textbook. Try problems both orally and in writing. Learn any rule or definitions.

Tuesday. Work more problems in Monday's section. Reteach anything that was difficult on Monday. Sometimes let the children make up problems of their own.

Wednesday. Proceed as on Monday.

Thursday. Proceed as on Tuesday.

Friday. Study a new section if your schedule requires taking more than two sections per week. Or continue work on the previous sections, if they are not finished yet.

Planning Guide for Grade 5

(Ray's Practical Arithmetic)

UNIT: NOTATION

Week 1

Sections 1 to 14

Objectives: to be able to read and write large numbers in the Arabic system and numbers to the thousands in the Roman system; and to understand

and use the vocabulary of number systems which is presented in this unit.

At the rate of three sections per day, this unit can be covered in a week. Some sections give practice in writing numbers, and others work well with the recitation method. Children can study the content and then try answering questions about what they studied. This gives them opportunity to use, as well as hear, the vocabulary which they need to learn thoroughly.

UNIT: ADDITION

Weeks 2 and 3

Sections 15 to 20

Objectives: to review the basic addition facts and to learn how to do addition problems with carrying.

Check up on your children's mastery of the basic addition tables. Plan extra practice for those who need it. Spend sufficient time on sections 18 and 19 that the children are able to do problems with carrying. Mastery of the tables and of carrying are so important that you should spend an extra week or two on this unit if necessary. The extra time can be made up on later units.

Use variety in your classroom procedure. Sometimes you can have the children work on the chalkboard, or on slates or paper. You or a child can read aloud a problem, then everyone works it. They can compare answers, and if they do not agree, figure out what is wrong. One child could work on a calculator and see if his answers are the same as others. Children need to learn that sometimes the answer obtained on a calculator is wrong, because sometimes a number may be entered incorrectly.

Use as many of the problems as you feel your children need for drill. It is not necessary to use all of them.

For long problems, some children may work out shortcuts to share with each other. One challenging shortcut is in section 20. Another you can teach, is to look for combinations of 10. For instance, in a column check off 6 and 4, and 7 and 3, and then add the remaining numbers to 20. Many variations can be worked on this. For instance, if there is not a 3 to go with the 7, but there is a 4, use the shortcut anyway. Simply start with 20 and 1 more. Some children profit from the mental challenge of thinking up better ways to add.

UNIT: SUBTRACTION

Weeks 4 and 5

Sections 21 to 26

Objectives: to review the basic subtraction facts and to learn how to do problems with borrowing.

Check to see that all your children know the basic facts in section 23. Plan extra practice for those who need it. Sections 25 and 26 show the method of borrowing. Be sure everyone learns that, too. Spend an extra week or two on this unit if necessary. For variety in the practice, let the children sometimes work on paper or slates, and sometimes at the chalkboard. Some children may use calculators and compare their answers with those who are working on paper.

Use as many problems as you feel your children need for drill. It is not necessary to use all of them.

UNIT: MULTIPLICATION

Weeks 6 to 8

Sections 27 to 34

Objectives: to review the multiplication tables through 12s and to be able to understand and compute multiplication problems with simple numbers.

Some sections in this unit look short, but each contains important concepts, so review them often. Make sure your children can explain the principles, rules, and vocabulary. Section 27 teaches that multiplication is a short way to add. Have the students think up more examples and do the problems both ways, as shown in the book example. Section 29—the tables—should be practiced every week of this unit to achieve speed, accuracy, and overlearning. Section 30—the principle of order—is important to understand. Have the children demonstrate this with checkers or other manipulative objects. Review the vocabulary in section 14 while studying this section.

Section 31 contains rules for both long and short multiplication, so stay in this section long enough for the children to become competent in this work. Each child can make up two problems each day for his classmates to work. Sometimes use calculators. Explain to the children that though calculators are easy and timesaving, it still is important to know in their heads what the calculators are doing.

Use as many of the problems in each section as you feel your children need for practice. It is not necessary to use all of them.

Sections 32 through 34 present special cases which can be used as shortcuts or sometimes to enable problems to be done mentally rather than on paper. Advanced students enjoy the challenge of this work. If you have used extra time in the units thus far, making sure your children master the basics, these three sections could be skipped to help you get back on schedule.

Use as many of the practice problems as your children need to become proficient in multiplying. It is not necessary to use them all.

UNIT: DIVISION

Weeks 9 to 13

Sections 35 to 49

Objectives: to review the division tables through 12s and to be able to understand and compute division problems with simple numbers.

At the rate of three sections per week, you can cover this unit in five weeks. But be a little flexible with that schedule, spending more time on the harder parts and less time on the easy parts. Practice often on the tables in section 38, to achieve speed, accuracy, and overlearning. Review often the rules and principles in all sections. If you prefer, you may teach the children to write the quotient above the dividend instead of to the right of it.

Use as many of the practice problems as needed to help your children become proficient in division. It is not necessary to use them all.

UNIT: COMPOUND NUMBERS

Weeks 14 to 20

Sections 50 to 82

Objectives: to be able to compute problems involving money, measures, and time.

It takes one section per day to cover this unit in

seven weeks, and you can use that plan for an advanced course, using as many problems in each section as you feel are necessary. For a basic course, choose some sections to omit. For example, you may omit or treat briefly section 61 on dry measures if you think your children will not have much use for knowledge of bushels and pecks. Or you may omit sections 81 and 82 on longitude and time if you think this is difficult for your particular group (though it is an extremely interesting section). By such means, you may gain time that you added in previous units to make sure your children mastered the basics.

You can supplement the study of time by teaching about standard time. This system was first introduced in 1883 by railways in the United States and Canada and has since been widely adopted in other countries. It uses standard meridians 15° apart, and a time zone is the area spreading to 7½° on either side of one of these meridians.

UNIT: FACTORING

Weeks 21 to 24

Sections 83 to 91

Objectives: to understand prime numbers and factors, and to use factors in computations.

Spending two days on each section will enable you to cover this unit in four weeks. You can teach a section on the first day and have the children "teach" it back to you on the second day. Advanced students with a good understanding of numbers may need less time to master this unit. For those children, provide enrichment by using some of the ideas in the game and project section of this manual.

Remember that it is not necessary for students to work all the practice problems. Also, adjust for students who have great difficulty with this topic, by moving on after a time to some sections of the fractions unit.

UNIT: FRACTIONS

Weeks 25 to 36

Sections 92 to 130

Objectives: to be able to add, subtract, multiply, and divide common fractions and mixed numbers.

Taking four sections per week, you can cover this unit in twelve weeks. There are rules and principles to learn and vocabulary to become familiar with, as well as problems to solve. At first, do the problems one at a time so the children can get immediate feedback on their work. Later, they may work groups of problems. Assign a reasonable number of problems, according to the time your students have available.

Remind the children how much they have learned this year. Have them look back through the book and see the many kinds of problems they were not able to do before but now are able to do. Work some of them for a review.

PROGRESS RECORD
Grade 5

Names	notation		add with carrying	subtract with borrowing	multiplication			division			compound numbers		factoring, find:			fractions	
	Arabic	Roman			tables	short	long	tables	short	long	tables	problems	prime factors	G.C.D.	L.C.M.	add and subtract	multiply and divide

Testing Fifth Graders

There are eight units to study during fifth grade. And a test for each unit is shown on the following pages of this manual. They can be used for pretests as well as posttests. When you are about to start the units on basic addition, subtraction, multiplication and division, you may want to give the tests as pretests to see what skills your children have in these operations. From the results of a pretest, you can determine how much time to spend on a unit. Any time gained on these early units can be used on later units, such as factoring and fractions.

More tests for each unit are found in *Ray's Test Examples.* Use selected problems or full tests from that book as you see need for them during your study of each unit.

Test Schedule

Unit 1: Notation. This test provides you with a good check-up on the children's knowledge of Arabic and Roman numerals and on vocabulary used when talking about numbers. Use it as a pretest or posttest or both.

Also see tests 11 and 12 (articles 11 and 12 in *Ray's Test Examples*).

Unit 2: Addition. This test has sections sampling the addition tables, problems without carrying, problems with carrying, and story problems. By noticing where your children make errors, you can tell what topics to teach or reteach. Use the test as a pretest or posttest or both.

Unit 3: Subtraction. This test has sections sampling the subtraction tables, problems without borrowing, problems with borrowing, story problems, and vocabulary associated with subtraction. Notice where your children make errors and you can focus your teaching where they need it most. Use the test as a pretest or posttest or both.

Unit 4: Multiplication. This test has sections on vocabulary, tables, problems with smaller numbers, and problems with larger numbers. Results of the test will tell you what your children have learned and what they may need reteaching on.

Unit 5: Division. This test has sections on vocabulary, tables, short division and long division. Results of the test will show whether your children have the important knowledge and skills of this unit.

Unit 6: Compound Numbers. This test samples the facts and skills in the unit on compound numbers. If you omit any parts of the unit with your class in order to stay closer to your planned schedule, be sure to adjust the test accordingly.

Unit 7: Factoring. This test shows your children's skill and understanding of factoring. This knowledge is important because it makes much other work in arithmetic easier. So require a rather high score for passing, and plan extra study to help everyone pass the test.

Unit 8: Fractions. The fractions test will show your children's knowledge and skills in all the common operations of fractions. By noticing which sections they make errors in, you can determine where they may need additional study.

Notation

Name ______________________________ Date _______________

Write these numbers in the Arabic system.

1. Three hundred three _______________
2. One thousand two hundred _________________
3. Two thousand one hundred _________________
4. One hundred thousand and ten _________________
5. Forty-five million eighty-three thousand and twenty-six.

____

6. Eighty billion seven hundred and three million five hundred and four ____

Write these as Roman numerals.

7. 1 to 10 ____
8. 11 to 20 ____
9. 30 __________ 10. 40 ________ 11. 50 ________
12. 100 __________ 13. 250 ________ 14. 500 ________
15. 1000 __________ 16. this year's date _________________

Write these words beside their definitions.

integer **abstract number** **concrete number** **compound number**

17. a whole number ___________________________________
18. a number applied to objects _________________________
19. a number without objects in mind ___________________
20. a number made up of two or more concrete numbers, such as 3 yards 2 feet ____

Score ____________

Addition

Name ______________________________ Date _______________

Tables:

2 + 9 = ________ 9 + 9 = ________ 7 + 8 = ________

8 + 6 = ________ 6 + 9 = ________ 8 + 5 = ________

Without carrying:

240	4321	50230
132	1254	3105
25	3120	423

With carrying:

184	103	5943
216	405	6427
135	764	8204

13. How many days are in October, November, and December all together? ________

14. General Washington was born in 1732 and lived 67 years. In what year did he die? ________

15. A man has 4 flocks of sheep. In the first are 65 sheep and 43 lambs, in the second are 187 sheep and 105 lambs; in the third, 370 sheep and 243 lambs; and in the fourth 416 sheep and 95 lambs. How many sheep does he have and how many lambs? ________

16. 23 + 41 + 74 + 83 + 55 = ________________

17. 11 + 22 + 33 = ________________

18. 263 + 104 + 38 = ________________

19. 51 + 76 + 4 = ________________

20. 7203 + 48 + 21 = ________________

Score ____________

Subtraction

Name ______________________________ Date _______________

Tables:

15 − 6 = ______	14 − 9 = ______	15 − 8 = ______
17 − 8 = ______	15 − 7 = ______	13 − 4 = ______
19 − 9 = ______	13 − 7 = ______	17 − 9 = ______

Without borrowing:

734	8752	79484	49528
531	3421	25163	16415

With borrowing:

7640	848	1940	4498
1234	57	780	2500

18. The answer in subtraction is called the _______________ or the _________________________.

19. Jason had 100 dollars in the bank and he withdrew 30 dollars. How much did he have left? ____________________

20. America declared her independence in 1776. How many years ago was that? ______________________________

21. Take 17 cents from 63 cents. ____________________

22. Subtract 423 from 569. __________________________

23. Seven hundred minus 550 is. ____________________

24. Forty from 63 leaves. __________________________

25. Start with 369 and take away 48. _______________

Score _____________

Multiplication

Name ______________________________ Date _______________

Vocabulary:

$$\begin{array}{r} 47 \\ \times\ 3 \\ \hline 141 \end{array}$$

1. Which number is the multiplier? _______________
2. Which number is the multiplicand? _________________
3. Which number is the product? _________________
4. If the 141 is apples, what other number is apples?

____

5. If the product is marbles, what other number is marbles?

____

6. Which number must be abstract? _________________
7. Which two numbers can be concrete numbers?

____

8. What means "multiplied by"? _________________

____

Tables:

12 × 12 = ______	11 × 9 = ______	10 × 11 = ______
11 × 8 = ______	10 × 12 = ______	11 × 12 = ______
11 × 11 = ______	6 × 12 = ______	12 × 9 = ______

Multipliers 12 or below:

231	132	644	1200	5661
2	3	12	9	10

Multipliers above 12:

235	869	346	425	485
13	19	21	364	526

28. Give a rule for multiplying by 100. ___________________

____

Score _____________

Division

Name ______________________________ Date _______________

Vocabulary:

$$3\overline{)18}\ \text{with quotient } 6$$

1. Which number is the divisor? _______________
2. Which number is the dividend? _______________
3. Which number is the quotient? _______________
4. Which number is the product of the other two? _______________

Tables:

132 ÷ 11 = __________	120 ÷ 10 = __________
121 ÷ 11 = __________	108 ÷ 12 = __________
96 ÷ 8 = __________	132 ÷ 12 = __________
144 ÷ 12 = __________	110 ÷ 11 = __________
99 ÷ 9 = __________	

Short division:

15. Divide 693 by 3.
16. Divide 4682 by 2.

Long division:

17. Divide 3465 by 15.
18. Divide 71104 by 88.
19. Divide 25312 by 112.
20. Divide 4700 by 100.

Score __________

Compound Numbers

Name ______________________ Date ____________

5 yards = 15 feet

1. When you change yards to feet, do you change the value of the number? ____________

2. Do you change the denomination of the number?

____

____

3. Nine dollars = ______________ cents.

4. Eight quarts = ______________ peck(s).

5. Eight quarts = ______________ gallon(s).

6. One mile = ______________ feet.

7. One square yard = ______________ square feet.

8. World War I began July 28, 1914 and World War II began September 3, 1939. How much time passed between them? ________

9. There are 14 pounds 4 ounces of beans. To put them in 4 containers equally, how much should you put in each container? ________

10. In one urn there are 2 gallons 2 quarts and 1 pint of punch. In another there are 1 gallon 3 quarts. How much punch is there all together? ________

11. Each crock pot holds 3 quarts 1 pint of chile and there are 6 pots at the chile supper. How much chile is there? ________

12. A farmer took 5 bags of wheat to the mill. Each bag contained 2 bushels 3 pecks of wheat. How much wheat did he take in all? ________

Score ________

Factoring

Name ______________________________ Date _______________

Vocabulary:

3 6

1. Which number is a multiple of 2? _________________
2. Which is a prime number? _________________
3. Which is a composite number? _________________
4. Which is a factor of 9? _________________
5. Which is a common factor of both numbers? ________

____

Find the prime factors of:

30	_____________	48	_____________	22	_____________
15	_____________	105	_____________	46	_____________

Find the greatest common divisor of:

16, 24, and 40	___________	75, 125, and 165	___________
78 and 130	___________	72, 120, and 132	___________

Find the least common multiple of:

4, 6, and 8	___________	6, 8, 9, and 12	___________
6, 10, and 15	___________	10, 12, 15, and 20	___________

20. What is a short way to divide 21 times 2 by 7 times 5?

21. What is a short way to divide 6 times 783 by 6 times 3?

Score _____________

Fractions

Name ______________________________ Date _______________

Vocabulary:

1. The upper number in a fraction is called the ____________________________.

2. The lower number in a fraction is called the ____________________________.

3. The number that expresses a part of one or more units is called a ____________________________.

4. A fraction of less than 1 is called a(n) ____________________________ fraction.

5. A fraction greater than 1 is called a(n) ____________________________ fraction.

Change to whole or mixed numbers:

$^{19}/_{7} =$ $\quad$ $^{12}/_{4} =$ $\quad$ $^{53}/_{4} =$ $\quad$ $^{75}/_{5} =$

Reduce to lowest terms:

$^{18}/_{30} =$ $\quad$ $^{12}/_{18} =$ $\quad$ $^{60}/_{90} =$ $\quad$ $^{126}/_{198} =$

Change to their least common denominator:

$^{1}/_{2}$ and $^{2}/_{3}$ _______________ $\quad$ $^{3}/_{8}$ and $^{4}/_{5}$ _______________

$^{4}/_{7}$ and $^{3}/_{10}$ _______________ $\quad$ $^{1}/_{2}$ and $^{5}/_{8}$ _______________

Solve:

$^{1}/_{4} + {}^{7}/_{8} + {}^{11}/_{12} =$ $\quad$ $2^{1}/_{2} + 3^{1}/_{3} =$

$^{3}/_{4} - {}^{2}/_{3} =$ $\quad$ $5^{2}/_{3} - 4^{3}/_{4} =$

$^{3}/_{4} \times 3 =$ $\quad$ $^{3}/_{4} \times {}^{5}/_{7} =$

$2^{2}/_{5} \div 6 =$ $\quad$ $4^{3}/_{4} \div 5^{1}/_{8} =$

Score _______________

Answer Key

NOTATION. 1) 303. **2)** 1200. **3)** 2100. **4)** 100,010. **5)** 45,083,026. **6)** 80,703,000,504. **7)** I, II, III, IV, V, VI, VII, VIII, IX, X. **8)** XI, XII, XIII, XIV, XV, XVI, XVII, XVIII, XIX, XX. **9)** XXX. **10)** XL. **11)** L. **12)** C. **13)** CCL. **14)** D. **15)** M. **16)** **17)** integer. **18)** concrete number. **19)** abstract number. **20)** compound number.

ADDITION. Tables, left to right: 11, 18, 15, 14, 15, 13. **Without carrying:** 397, 8695, 53758. **With carrying:** 535, 1272, 20574. **13)** 92. **14)** 1799. **15)** 1038 sheep and 486 lambs. **16)** 237. **17)** 66. **18)** 405. **19)** 131. **20)** 7272.

SUBTRACTION. Tables, left to right: 9, 5, 7, 9, 8, 9, 10, 6, 8. **Without borrowing:** 203, 5331, 54321, 33113. **With borrowing:** 6406, 791, 1160, 1998. **18)** difference or remainder. **19)** 70 dollars. **20)** . **21)** 46 cents. **22)** 146. **23)** 150. **24)** 23. **25)** 321.

MULTIPLICATION. 1) 3. **2)** 47. **3)** 141. **4)** 47. **5)** 47. **6)** 3. **7)** 47 and 141. **8)** x. **Tables, left to right:** 144, 99, 110, 88, 120, 132, 121, 72, 108. **Multipliers 12 or below:** 462, 396, 7728, 10800, 56610. **Multipliers above 12:** 3055; 16,511; 7266; 154,700; 255,110. **28)** Annex two zeros to the multiplicand.

DIVISION. 1) 3. **2)** 18. **3)** 6. **4)** 18. **Tables, Column 1:** 12, 12, 12, 12, 11. **Column 2:** 12, 9, 11, 10. **15)** 231. **16)** 2341. **17)** 231. **18)** 808. **19)** 226. **20)** 47.

COMPOUND NUMBERS. 1) no. **2)** yes. **3)** 900. **4)** 1. **5)** 2. **6)** 5280. **7)** 9. **8)** 25 years, 1 month, 6 days. **9)** 3 pounds, 9 ounces. **10)** 4 gallons, 1 quart, 1 pint. **11)** 5 gallons, 1 quart. **12)** 13 bushels, 3 pecks.

FACTORING. 1) 6. **2)** 3. **3)** 6. **4)** 3. **5)** 3. **Prime factors, left to right:** 2, 3, and 5; 2, 2, 2, 2, and 3; 2 and 11; 3 and 5; 3, 5, and 7; 2 and 23. **Greatest common divisors:** 8, 5, 26, 12. **Least**

common multiples: 24, 72, 30, 60. **20)** Cancel the 7s. That leaves $3 \times 2 - 5$, which is $6 - 5$. **21)** Cancel the sixes. That leaves $783 - 3$.

FRACTIONS. 1) numerator. **2)** denominator. **3)** fraction. **4)** proper. **5)** improper. **Whole or mixed numbers:** $2\frac{5}{7}$, 3, $13\frac{1}{4}$, 15. **Lowest terms:** $\frac{3}{5}$, $\frac{2}{3}$, $\frac{2}{3}$, $\frac{7}{11}$. **Least common denominators:** $\frac{3}{6}$ and $\frac{4}{6}$, $\frac{15}{40}$ and $\frac{32}{40}$, $\frac{40}{70}$ and $\frac{21}{70}$, $\frac{4}{8}$ and $\frac{5}{8}$. **Solve, left to right:** $2\frac{1}{24}$; $5\frac{5}{6}$; $\frac{1}{12}$; $\frac{11}{12}$; $2\frac{1}{4}$; $\frac{15}{28}$; $\frac{2}{5}$; $\frac{38}{41}$.

Teaching Sixth Grade

Using *Ray's New Practical Arithmetic*

page

Teaching Sixth Grade

The main objectives of the sixth grade curriculum are to be able to use decimal fractions and percentages in solving problems of discount, interest, and other business applications; to know ratio, proportion, powers, square and cube roots, arithmetical and geometrical progression, the metric system, and the geometry of plane and solid figures. This covers sections 131 to 270 in *Ray's Practical Arithmetic.*

Use group teaching methods that will enable you to see how the children are progressing. Study the examples and rules directly from the textbook. Then you or a student can read problems one at a time for the group to work. If several children work at the chalkboard, you can watch their procedures and children can explain to the group how they solved the problems.

This year your children will study several topics or concepts of practical use in everyday money matters. Among them are: discounts, interest, commissions, profit and loss, and mortgages. Use these opportunities to study real life situations. Newspapers are good source materials for some of these topics. The financial pages, bank ads, and sale ads are useful study materials. Often you and the children themselves can make up problems from these materials to supplement those in the textbook. In home schools you can study family insurance policies, mortgages, savings, investments, installment loans, taxes, and other money matters. In classrooms you might also take advantage of these source materials by suggesting homework on these with cooperation from the parents. But respect family privacy and do not have any of the information brought back to class.

These practical topics consist of so much learning that you may find you are not able to complete all the work this year with your children. If this is the case, do not worry too much about it, as the seventh and eighth grade courses allow time for studying these topics in greater depth, or for studying topics you have to skip this year, or for reteaching any topics you think your children do not understand well.

An important principle to remember in teaching is that several exposures are often needed for children to master new material. You can't assume that if children seem to understand a new topic and do the problems one day that they will remember how it all works on the next day. Reteaching, review, and many opportunities for practice and discussion should be planned into every unit. If we stop to notice, we see that we adults are the same way in learning something new and difficult. Once through is not enough. This insight can help us have more patience with the children.

Use the following schedules and guides to help plan your year's work with the children. Be flexible in following the guides, adjusting as needed for the children you teach. Help your children to a good understanding of decimal fractions and percentages. That is basic. If you have to skip something, don't skip those basics, but skip, instead, some of the applications of those skills, such as exchange, insurance, or taxes.

Be enthusiastic. Learn to like arithmetic better than ever before, and watch the happy results in your students. Have a happy year.

A Typical Day

A typical day in sixth grade can often start with discussion about something from the previous day's lesson. Children can be encouraged to bring sample problems of their own. They can bring newspaper ads or articles of interest. They can tell about life situations where they used some of the skills they are learning in class.

After such friendly and motivating discussion, the class can work on the textbook lesson for the day. This will take the major portion of the period. Guide the work yourself, as this method has been found to be most effective for good learning. After a new topic is learned and you think the children can do the problems on their own, then they may work a few problems individually. This will help each child see how he can proceed on his own, and whether he needs help on any part of the process.

Some days, you can individualize by making different assignments. Have remedial or catch-up work for those who need it and extra-challenge work for those who don't. Use ideas from the projects and games sections of this manual.

A Typical Week's Schedule

Monday. Study a new section in the textbook and try some of the problems.

Tuesday. Work more problems of the kind studied on Monday. Individual students may demonstrate their work on the chalkboard and explain the process. Review the rules which apply.

Wednesday. Study a new section in the textbook. See how it is like the previous work and how it is different. Learn any rules or definitions and work some problems.

Thursday. Review Wednesday's topic. Talk about any real life examples the students have brought. Begin another new section in the textbook.

Friday. Reteach and work more on the section begun on Thursday. Play an arithmetic game or work on a project.

Planning Guide for Grade 6

(Ray's Practical Arithmetic)

UNIT: DECIMAL FRACTIONS

Weeks 1 to 4	**Sections 131 to 188** *Objectives:* to be able to read, write, and compute

decimal fractions; and to understand the orders and their relation to common fractions.

At the rate of six sections per week, you can cover this unit in four weeks. But since the sections vary in size, you may adjust that schedule slightly and still be able to finish. For instance, the first week you may only cover five sections because 135 is long. Then on the second week, cover seven sections. Do not assign more practice problems than the children are able to compute in their allotted time.

Beginning with section 143, spend some time each day on estimating answers. Have the children look at the highest orders—the left-hand figures—in their problems and name a number that should be close to the actual answer. Then they can work the problems on paper or sometimes on calculators and check answers against their estimations. Children should learn that a misplaced decimal point or some other error that is easy to make can throw an answer way off. They also should learn not to blindly depend on calculators or computers being right.

UNIT: THE METRIC SYSTEM

Weeks 5 to 6

Sections 155 to 161

Objectives: to know the measures of length, area, capacity, and weight; to be able to compute problems in these systems; and to convert from each system to the other.

In working these sections, children will discover that it is easy to compute within the metric system because everything is in tens, but converting between metric and other systems is more difficult. Compare a meter and a yard to find which has more prime factors (section 85). Do the same with a foot and a decimeter. Let the children discuss which system they prefer to use.

Besides using some of the problems and tables from the book, provide actual experience in weighing, measuring and pacing off the distances. Also, teach the meanings of prefixes used in the metric system. Children may check a dictionary and think of other words they know which use the prefixes, such as century or millennium. *Deci* and *deka* mean 10 or 10ths; *centi* and *hekto* mean 100 or 100ths; and *milli* and *kilo* mean 1000 or 1000ths.

UNIT: PERCENTAGE

Weeks 7 to 11

Sections 162 to 180

Objectives: to understand and compute bases, rates, and percentages; and to apply this knowledge to commissions, discounts and other business uses.

For a basic course, have the children master the sections through 169, using as many drill problems in each as you feel necessary. For an advanced course, work on through some or all of the business applications. At the rate of four sections per week you can cover the unit in five weeks' time.

UNIT: INTEREST

Weeks 12 to 16

Sections 181 to 193

Objectives: to understand the concepts of principal, simple interest, and compound interest; and to be able to compute problems involving them.

By spending two days on each section, you can cover the unit in five weeks' time. Instead of using the table on page 220, have the children search the business and real estate sections of newspapers to learn about current interest rates. What are mortgage rates now in your area? What is the T-bill rate this week? Can the children find out what is meant by prime rate, discount rate, and broker loan rate on money? What other information and questions can they get from newspapers?

During study of other sections, also try to make use of current situations. Particularly in home schools, children may learn about family loans where interest is paid and family accounts where interest is earned. Let them do some computing on these. Can they think of ways to earn more interest or to pay less interest?

Remember to be selective in the problems you assign. Do not overburden your students with work.

UNIT: DISCOUNT

Weeks 17 to 18

Sections 194 to 199

Objectives: to understand that discount on loans is interest paid in advance, and to be able to find discounts and face amounts when the other information is given.

Taking three sections per week, you can cover this unit in two weeks. Supplement the unit with current information, if you can. For instance, from a bank which handles bonds, obtain a list of bonds now available. Study the list to see whether bonds are selling at a discount or at a premium. Make up some problems from the list. What is the discount on a certain bond? What is the largest discount on the list? Which bond would the children like to buy?

If your students meet the objectives stated above by such current life problems, feel free to skip drill problems in these sections. This topic and other following topics are studied again in seventh and eight grades, so if your children achieve only introductory understandings this year your job is well done.

UNIT: EXCHANGE

Week 19

Sections 200 to 202

Objective: to understand the principle of exchanging money within a country or between countries.

A simple form of draft is a check. All children can learn how to write checks. They may also learn about shillings, francs and marks and the idea of exchanging dollars for them. Advanced students may learn of current rates of foreign exchange from the business section of many newspapers. Work some problems from the newspaper, finding, for instance, how much U.S. money it takes to buy a Japanese yen.

UNIT: INSURANCE

Week 19

Sections 203 to 205

Objective: to understand the concept of insurance and some of the vocabulary related to it.

The work in this unit will teach some basic principles of insurance. You may supplement it with information or problems about a family policy. For a basic course, this unit may be omitted.

UNIT: TAXES

Weeks 21 to 22

Sections 206 to 212

Objective: to understand the concept of taxation.

The first week study four sections, and the second week, three. In section 210 you may supplement by teaching also about today's direct tax—the income tax. For a basic course, this unit may be omitted.

UNIT: RATIO

Week 23

Sections 213 to 220

Objective: to understand the concept of ratio and to be able to compute problems involving ratio.

For a basic course you may go as far as section 216, and for an advanced course continue with the other sections.

UNIT: PROPORTION

Weeks 24 to 27

Sections 221 to 232

Objective: to understand the concepts of proportions, shares, and averages and to be able to work problems involving them.

The form "2 is to 4 as 3 is to ?" is known to every test taker. This unit on proportions will help children understand such statements and to find any missing term in them. This skill is really quite useful in real life as well as in test taking, as your children will see from the problems in this unit.

At three sections per week, you can cover the whole

unit in four weeks. For a basic course, help the children master at least through section 224. Then add as many more sections as they have time for.

UNIT: INVOLUTION AND EVOLUTION

Weeks 28 to 31

Sections 233 to 245

Objectives: to be able to find squares and higher powers of common numbers, and to find square and cube roots of common numbers.

At three sections per week, you can cover this unit in four weeks. Emphasize understanding the concepts, which is best done by working with smaller numbers and with the geometric drawings. For a basic course, you can do sections 233 through 235, the last paragraph of page 304, and sections 239 and 240.

UNIT: GEOMETRY

Weeks 32 to 35

Sections 246 to 263

Objectives: to be able to find lengths, areas, and volumes of common plane and solid figures.

At approximately one section per day you can complete this unit in four weeks. But be flexible with this plan. For instance, on the first day the children may study the definitions in the first section and work some of the problems in the next section. Then on the second day they may review the definitions and work more of the problems.

While working on this unit, let the children have some fun with geometric designs, as suggested in the project section of this manual.

UNIT: PROGRESSIONS

Week 36

Sections 264 to 270

Objectives: to understand and compute problems involving arithmetical and geometrical progression.

This is another topic used sometimes for test questions and tricky math puzzles. Experience with progressions is good for the children's thinking skills. It will take a little more than one section per day to finish this unit in a week. This unit especially fits an advanced course, and may be omitted in a basic course.

Encourage the children by helping them see how much they have learned this year. They may leaf through the book or look at the table of contents and reminisce about their studies during the year. What were their favorite topics? Which were easiest or most fun? Which were most difficult? Which are they the proudest about knowing?

PROGRESS RECORD
Grade 6

Names	decimal fractions				metric		percentage			ratio and propor-tion	roots and powers	geometry	progres-sions
	read and write	convert to common fractions	add, subtract, multiply, divide	compound numbers	know tables	solve problems	know equivalent fractions	know definitions	solve interest, discount, etc.				

Testing Sixth Graders

How should you use tests? One of the most important uses is as a pretest before a unit. The purpose of pretests is to see what the children may already know about the new work. Such a diagnosis helps you plan the children's work. You can spend time where it is needed the most.

A second common use of tests is at the end of a unit. Handle this use in a way that will not make children feel like failures. First, teach the unit well, so they know the material and can be successful on the test you give them. Second, stress how much they have learned instead of how much they miss. Use the missed problems to plan a review. The children themselves may help you figure out why they missed problems and what is needed to remedy the situation. End-of-unit tests also act as motivators. Children study harder because a test is coming up.

Tests are also for any other use that may fit your teaching plans. One idea is a practice test before the "real thing" so children can build confidence or see what to study before the real test. Another idea is to diagnose at any time during a unit that you feel a need for it.

By all these means, tests contribute to learning. Use them to achieve good learning in your students. Avoid using them as heavy weights that put too much stress on children and make some of them feel that they are failures in life. Use tests so you can teach better and so your children can learn better.

The following pages show ready-made tests for each sixth grade unit. You may easily make additional tests by selecting problems from matching article numbers in *Ray's Test Examples*.

Testing Schedule

Unit 1: Decimal Fractions. Use this test either as a pretest or a posttest, or both.

Unit 2: The Metric System. This test is a checkup on general knowledge about metric measurements. You may use it as both a pretest and a posttest.

Unit 3: Percentage. This test includes per cent and common fraction equivalents and story problems. Together these will show whether your children have a basic understanding of percentages and can handle computations with them.

Unit 4: Interest. This test includes some rather long and complex problems, so your children may need extra paper to do their calculations on. They can show on this test whether they can handle the concept of interest in some common life situations. You may want to look over the test before you begin teaching the unit so you can see what kinds of skills the children will need in order to pass it. Use the test after the unit.

Units 5, 6, 7, and 8: Discount, Exchange, Insurance, and Taxes. This test covers a variety of applications of per cent and decimals, and you may use it after the children study these four short units.

Units 9 and 10: Ratio and Proportion. The two short units on ratio and proportion are closely related in content, so you may use this test after both units are studied.

Units 11 and 12: Involution and Evolution. The test supplied for these units will fit the basic course suggested in the planning guide. If you have students doing advanced work, you may add more problems to make a test for them.

Unit 13: Geometry (or Mensuration). This test provides you with a check-up on your children's ability to find areas and volumes of common shapes.

Unit 14: Progressions. On this test, your students can show whether they understand and can apply major rules for working with arithmetical and geometrical progressions.

Decimal Fractions

Name ______________________________ Date _______________

Write these as decimal fractions.

$\frac{1}{10}$ =	$\frac{1}{100}$ =	$\frac{6}{10}$ =
$\frac{2}{100}$ =	$\frac{25}{100}$ =	$\frac{1}{1000}$ =

Write these as common fractions.

.9 =	.28 =	.651 =

Write these as decimal fractions.

Seven tenths ___________________________

Fifty-four thousandths ___________________________

Seventy-five hundredths ___________________________

.7 .70 .07 .007

13. Which number(s) has the greatest value? _______________
14. Which number(s) has the least value? _______________
15. Which two numbers have the same value? _______________
16. Which number means the same as $\frac{7}{100}$? _______________

---

17. Add 37.25, 1.113, and 130.6. _______________
18. Subtract 20.15 from 100. _______________
19. Multiply 748 by .25. _______________
20. Divide 1.125 by .03. _______________

Score _____________

The Metric System

Name ______________________________ Date ______________

Vocabulary matching:

1. a measure of length — liter
2. a measure of capacity — gram
3. a measure of weight — meter

4. A kilometer is closest to a (foot, yard, mile).

5. A meter is closest to a (foot, yard, mile).

6. To change units to a higher denomination in the metric system, you move the decimal to the (left, right).

7. To change units to a lower denomination in the metric system, you move the decimal to the (left, right).

8. Foot races are sometimes measured in (millimeters, kilometers, milligrams).

9. Medicine is sometimes measured in (millimeters, kilometers, milligrams).

10. Gasoline is likely to be measured in (meters, liters, kilograms).

1 meter = __________ decimeters = __________ centimeters

1 gram = __________ decigrams = __________ centigrams = ____________ milligrams

__________ meters = 1 kilometer

__________ grams = 1 kilogram

15. One meter is 39.37 inches long. How long is 5 meters expressed in yards, feet and inches? ______________________

____

16. How long is a 50-yard dash if it is expressed in meters?

____

Score ____________

Percentage

Name ______________________________ Date ______________

$\frac{1}{100}$ is __________% $\frac{2}{100}$ is __________%

$\frac{5}{100}$ is __________% $\frac{50}{100}$ is __________%

Find 1% of 278. __________ Find 3% of 97. __________

Find 5% of 118. __________ Find 12½% of 200. ______

9. A man invested $3000 and made 15% profit. How much profit did he make? __________

10. Another man invested $3000 and made $360 profit. What per cent did he gain? __________

11. A third man bragged that he made a 20% gain of $400. How much did he invest in the beginning? __________

12. A salesman receives 5% commission on the goods he sells. After selling $4530 worth of goods, how much commission will he receive? __________

13. A jacket with an original price tag of $45 is on a sale rack with a sign announcing "30% off." How much would you have to pay for the jacket if you buy it on the sale? __________

14. If you pay $52 for a share of stock and it goes up to $65, you could sell the stock for what per cent gain? __________

15. The $52 stock paid a yearly dividend of $5.20. What per cent yield was that? __________

16. After the stock rose to $65 it was still paying a dividend of $5.20. What per cent yield was it then? __________

Score __________

Interest

Name ______________________ Date ______________

1. If your father lends you $100 to start a summer business, the $100 is called the (interest, principal, promissory note). ________

2. If your business works out well, you would like to repay your father three months later. At an annual rate of 12%, how much money would you need to repay both the principal and the interest? ________

3. If your business makes $240 profit you may want to put it in a bank. If the bank pays 6% interest and compounds it semiannually, how much interest would it earn in a year's time? ________

4. If the bank compounded the interest quarterly, how much interest would it earn in a year's time? ________

5. Would you rather have that bank compound the interest monthly or daily? ______________________ Why?
____

6. If your father owes $1800 on his car, and he is paying 16% interest, how much interest will he pay this month? ________

7. If your family owes $32,000 on the home mortgage and the interest rate is 13%, how much interest is due this month? ________

Score ________

Discount, Exchange, Insurance, and Taxes

Name ______________________________ Date ______________

1. A $1000 no coupon (non-interest bearing) bond matures in 3 years. It now sells for $700. What is the dollar amount of the discount and the per cent of discount? __________ __________

2. Mr. Moto lent a businessman $500 for 3 months. He discounted the interest, which was at an annual rate of 12%. What were the proceeds of the loan to the businessman? __________

3. What is the cost of a sight draft for $1500 to which ½% premium is added? __________

4. If the Canadian dollar is quoted at $.90 against the American dollar, how many Canadian dollars could you get in exchange for $25 American money? __________

5. What is the annual cost of insuring a $200,000 building for ¾ of its value at a premium of 1½%? __________

6. What must be paid yearly for $6000 of life insurance, the premium being $32.18 per $1000. __________

7. A father, at the age of 50, insures his life for $10,000 at $47.18 per $1000 annually. His son, aged 21, insures his life for the same sum, at an annual rate of $19.89. If each lives to the age of 71, which will pay the greater sum for insurance, and how much more than the other will he pay? __________ __________

8. A family's home is valued by the county at $80,000. The county tax rate is 6 mills, the water district rate is .6 mill, and the tax for public transportation is .2 mill. How much is the family's tax this year? __________

Score __________

Ratio and Proportion

Name ______________________________ Date _______________

What is the ratio of:

140 to 7 _______________ 30 to 100 _________________

56 to $2\frac{1}{3}$ ____________ 4.2 to .07 _________________

2 to 8 _______________ 3 weeks to 9 hours ______

Reduce the following to their lowest terms.

80 : 64 ___________ 8 oz. : 3 lb. _____________

77 : 33 ___________ 2 wk. : 2 da. ___________

48 : 36 ___________ 20 : 100 _______________

Find the missing terms.

3 : 6 :: 1: ____ 50 : 100 :: ____ : 40

9 : 12 :: 51 : ____ 12 : 16:: ____ : 28

8 : 9 :: ____ : 72 $\frac{1}{3}$: 2 :: ____ : 12

19. A watch lost $1\frac{1}{2}$ minutes in 2 days. How much time will it lose each week? ___________

20. During one week the class attendance was 27, 30, 30, 29, and 28. What was the average attendance that week? ___________

Score _____________

Involution and Evolution

Name ______________________________ Date _______________

$3^2 = 9$

1. In the figures above, the exponent is ___________.

2. The first power is ___________.

3. The square, or second power, is ___________.

Find the square, or second power, of:

10 ___________ 5 ___________

16 ___________ 213 ___________

Find the cube, or third power of:

4 ___________ 27 ___________

Find the fifth power of 7. ___________

Find the square root of:

144 ________ 49 ________ 625 ________

14. A ladder 53 feet long reaches a window 45 feet high. How far is the foot of the ladder from the bottom of the wall? __________

15. A mast 77 feet high stands in the middle of a ship 72 feet wide. How long is the rope ladder from the ship's side to the masthead? __________

Score ___________

Geometry

Name ______________________ Date ____________

Find the area of:

1. A floor 18 ft. 9 in. long and 13 ft. 4 in. wide. ________

2. A roof, each side being 16 ft. 6 in. by 40 ft. ________

3. A triangle 40 ft. long and 20 ft. high. ________

4. A circle whose radius is 25 ft. ________

5. The surface of a rectangular prism 4 ft. high, standing on a base 2 ft. square. ________

Find the volume of:

6. A prism 10 in. high, on a base 2 in. square. ________

7. A pyramid 48 ft. high, on a base 18 in. square. ________

8. A planet (sphere-shaped) whose diameter is 2240 miles. ________

9. A round swimming pool area 50 sq. yd. in diameter is surrounded by a fence 6 ft. high. At $.40 per sq. yd., what will it cost to paint the fence both inside and out? ________

10. A museum room is 70 ft. long, 40 ft. wide, and 22 ft. high. A skylight is 14 sq. ft., four doors are 5 ft. 3 in. by 10 ft., and 12 windows are 6 ft. by 9 ft. How many cans of paint should be bought to paint the walls and ceiling if each can covers 400 sq. ft.? ________

Score ________

Progressions

Name ______________________ Date ____________

1. The houses on an avenue are numbered regularly, beginning with 720. What will be the number of the 63rd house? ________

2. If lots 20 feet wide are numbered in regular order along a line, how far will the 36th lot be from the starting point? ________

3. A series of 10 terms begins with 7 and ends with 70. What is the common difference? ________

4. If you must count from 1 to 1000, using only 28 numbers, what will be the common difference between the numbers in your counting series? ________

5. Show a short way to sum up all the numbers from 1 to 10. ________

6. What is the sum of all numbers from 1 to 50? ________

7. A man pays you 5 cents for working one day, and each day he doubles your pay. How much will you earn on the tenth day? How much will you earn on the eleventh day? ________ ________

8. Another man pays you 5 cents for working one day, and each day he raises your pay by $1.00. How much will you earn on the tenth day? On the eleventh day? ________ ________

Score ________

Answer Key

DECIMAL FRACTIONS. Decimal fractions, Row 1: .1, .01, .6. **Row 2:** .02, .25, .001. **Common fractions:** $\frac{9}{10}$, $\frac{28}{100}$ or $\frac{7}{25}$, $\frac{651}{1000}$. **Decimal fractions:** .7, .054, .75. **13)** .7 and .70. **14)** .007. **15)** .7 and .70. **16)** .07. **17)** 168.963. **18)** 79.85. **19)** 187. **20)** 37.5.

THE METRIC SYSTEM. 1) meter. **2)** liter. **3)** gram. **4)** mile. **5)** yard. **6)** left. **7)** right. **8)** kilometers. **9)** milligrams. **10)** liters. **11)** 10, 100. **12)** 10, 100, 1000. **13)** 1000. **14)** 1000. **15)** 5 yards, 1 foot, 4.85 inches. **16)** 45.74+ meters.

PERCENTAGE. Row 1: 1%, 2%. **Row 2:** 5%, 50%. **Row 3:** 2.78, 2.91. **Row 4:** 5.9, 25. **9)** $450. **10)** 12%. **11)** $80. **12)** $226.50. **13)** $31.50. **14)** 25%. **15)** 10%. **16)** 8%.

INTEREST. 1) principle. **2)** $103. **3)** $14.62. **4)** $14.73. **5)** daily; The more often interest is compounded, the faster it grows. **6)** $24. **7)** $346.67.

DISCOUNT, EXCHANGE, INSURANCE, AND TAXES. 1) $300; 30%. **2)** $485. **3)** $1507.50. **4)** $22.50 or 22½. **5)** $2250. **6)** $193.08. **7)** son; $37.20. **8)** $544.

RATIO AND PROPORTION. Ratios, left to right: 20, $\frac{3}{10}$ or .3, 24, 60, ¼, 56. **Reduce:** 5:4, 1:6, 7:3, 7:1, 4:3, 1:5. **Missing terms:** 2, 20, 68, 21, 64, 2. **19)** 5¼ min. **20)** 28.8.

INVOLUTION AND EVOLUTION. 1) 2. **2)** 3. **3)** 9. **Squares:** 100; 25; 256; 45,369. **Cubes:** 64; 19,683. **Fifth power:** 16,807. **Square roots:** 12, 7, 25. **14)** 28. **15)** 85.

GEOMETRY. 1) 250 sq. ft. **2)** 146 sq. yd. 6 sq. ft. or 2780 sq. ft. **3)** 400 sq. ft. **4)** 19,635 sq. ft. **5)** 40 sq. ft. **6)** 40 cu. in. **7)** 36 cu. ft. **8)** 5,884,962,406.4 cu. mi. **9)** $251.33. **10)** 17 cans.

PROGRESSIONS. 1) 782. **2)** 700 ft. **3)** 7. **4)** 37. **5)** $[(1+10)\times 10]\div 2=55$. **6)** 1275. **7)** $25.60; $51.20. **8)** $9.05; $10.05.

Teaching Seventh Grade

Using *Ray's New Higher Arithmetic*

page

Teaching Seventh Grade

As you teach seventh graders, you will want to take each student as far as he can go. There is great variation in the skills and knowledge of seventh graders, greater than at any lower grade. This is normal and is a result of good teaching.

So your first job is to know as well as you can what your students are able to do. One way to achieve this is to give a pretest before each chapter. Choose a few representative problems and some of the vocabulary used in the chapter and make a test from them. From the results of the test, you can decide how to approach the study of that chapter.

Some students at this level need another chance to master the basics of arithmetic. Give them that chance. This time they may finally understand in a way they never could before. Other students will profit from an advanced look at the basics, such as *Ray's Higher Arithmetic* gives them. The principles, the extensive vocabulary, and the shortcuts to some of the operations all make enjoyable learning for such students.

For advanced students you can spend more time with topics such as circulating decimals, ratio, and proportion. With other students you can trim down the time on those topics and spend the time, instead, on more basic studies.

Below are two planning guides, one for a basic seventh grade course, and one for an advanced seventh grade course. Notice how many chapters are covered in each of these plans. Then adjust your plan during the year according to the ability of your own students. It may be that your plan will fall somewhere between the two which are shown.

Basic Course

(Ray's Higher Arithmetic)

Study chapters II through IX to review and increase ability

in the four basic operations with whole numbers, fractions, and decimals. Also study money and some of the measures in chapter XI.

Advanced Course

(Ray's Higher Arithmetic)

Study chapters I through XIII. Use pretests to help you determine how much time to spend on each of the topics. A few students may have time to take up one or more topics they find in the latter half of the book.

The following pages show ready-made tests for each seventh grade unit. You have the publisher's permission to make copies of the tests if your are using Ray's Arithmetics.

Introduction

Name ______________________________ Date _______________

True or False

__________ Arithmetic is a branch of mathematics.

__________ Mathematics is a branch of arithmetic.

__________ The solution is the answer to a problem.

__________ The solution shows how an answer was obtained.

---

5. When a rule is given in symbols instead of words, it is called a (formula, unit).

6. One thing is a (number, unit).

To show how many eggs are in
five dozen, you could write:
$5 \times 12 = 60$

7. In the above solution, 5 is (an abstract, a concrete) number.

8. Twelve is (an abstract, a concrete) number.

9. Sixty is (an abstract, a concrete) number.

---

10. Write at least two arithmetic signs that tell you to do an operation with numbers. ______________________________
____

11. Write at least two signs that show relationships between numbers. ______________________________

Score _____________

Numeration

Name ____________________________ Date ______________

581,342,621,473

1. Each group of three digits above is called a (unit, period, billion).

2. The first three digits tell how many (billions, trillions, quadrillions).

3. The 621 tells how many (hundreds, billions, thousands).

4. What digit tells how many tens of thousands there are? __________

5. What digit tells how many tens of millions? __________

6. What digit tells how many tens of billions? __________

Write these numbers.

7. four hundred twenty-eight thousand, six hundred forty-two ____________________

8. five hundred thousand sixty ____________________

9. eighty million six hundred twenty-three thousand ____________________

10. eight hundred million ____________________

Write these in Roman numerals.

11. From 9 to 12 ____________________

12. From 85 to 90 ____________________

13. Hundreds, from 300 to 800 ____________________

14. Write 900 in two different ways ____________________

Score __________

Addition and Subtraction

Name ______________________________ Date _______________

1. 622 + 387 + 494 + 365 + 938 = ____________

2. 7467 + 987 + 9816 + 42 = ________________

3. 453 + 268 + 2000 − 983 = ________________

4. 11,111 − 2844 − 2557 + 129 = ______________

5. A man had $1807 in the bank and wrote a check for $250. What was his balance after the check was paid? __________

6. A company owed Mr. Able $3800. Mr. Able bought from them $21,000 worth of goods, paying $5000 in cash and giving a note for the balance due. What was the amount of the note? __________

7. At a local election Mr. Brown received 1013 votes and Mr. Greene received 834 votes. By how many votes did Mr. Brown win? __________

8. Mr. Blue had $2125 in his bank account. He deposited $267 more, and then drew out $415 and $290. How much was left in the bank? __________

9. A publisher had 40,000 copies of a book printed, and sold 37,036. How many remain unsold? __________

10. From 17 million 14 thousand take 9 million 99 thousand 900. __________

Score __________

Multiplication and Division

Name ______________________________ Date _______________

1. $3823 \times 4 =$ ____________________

2. $7198 \times 216 =$ ____________________

3. $7575 \times 8600 =$ ____________________

4. Show a shortcut for multiplying 6413 by 98.

5. Using factors of the multiplier, show a shortcut for multiplying 4936 by 3208.

6. Divide 4865 by 5. ____________

7. Divide 382 by 17. ____________

8. Show an easy way to divide 4524 by 24.

9. Show an easy way to divide 68,475 by 25.

10. What number must be divided by 73 to make the quotient equal to the sum of 456 and 2893? ________________________

Score ____________

Properties of Numbers

Name ______________________________ Date _______________

Vocabulary for talking about numbers:

mixed number **prime number** **composite number**
integer **divisor** **multiple**

1. Which word means a whole number? _______________
2. Which means a whole number and a fraction?

____

3. Which cannot be divided by any number but itself?

____

4. Which has two or more factors?

____

5. Which refers to a product of a number and another number?

____

6. Which names a number that will exactly divide another number? _______________________________________

7. Which can be divided evenly by itself and by other numbers also? _______________________________________

Find the prime factors of:

45 72 98

Find the greatest common divisor of:

90 and 150 16, 40, and 88 42 and 70

Find the least common multiple of:

6, 9, and 12 10, 15, 20, and 30 70 and 105

Work each problem below and prove your answer by casting out nines.

$$\begin{array}{r} 5639 \\ +\ 481 \\ \hline \end{array} \qquad \begin{array}{r} 875 \\ -749 \\ \hline \end{array} \qquad \begin{array}{r} 78 \\ \times\ 8 \\ \hline \end{array} \qquad 2\,\overline{)\,498}$$

Score ___________

Common Fractions

Name ______________________________ Date _______________

$\frac{5}{9}$

1. Which numeral above is the numerator? _____________
2. Which is the denominator? _______________
3. This is called a (proper, improper) fraction.

____

Reduce each fraction to its lowest terms.

$\frac{42}{189} =$ $\frac{30}{45} =$ $\frac{156}{221} =$

Change these fractions to higher terms.

$\frac{3}{11} = \frac{\ }{99}$ $\frac{4}{9} = \frac{\ }{63}$ $\frac{3}{4} = \frac{\ }{100}$

Add.

$\frac{1}{6}$, $\frac{2}{9}$, and $\frac{5}{12}$ $2\frac{1}{4}$, $3\frac{2}{7}$, and $4\frac{5}{6}$

Subtract.

$\frac{9}{55} - \frac{1}{15}$ $\frac{4}{5} - \frac{5}{12}$

Multiply.

$\frac{10}{13} \times 12$ $8 \times \frac{3}{5}$

$6\frac{2}{3} \times 4\frac{1}{2}$ $32 \times 2\frac{3}{8}$

Divide.

$\frac{9}{16} \div 3$ $6 \div \frac{2}{3}$

$\frac{3}{4} \div \frac{1}{2}$ $\frac{21}{25} \div \frac{14}{15}$

Score _____________

Decimal Fractions

Name ______________________ Date ____________

Write each as a decimal fraction.

Fifty-one thousandths ____________ $\frac{7}{100}$ ____________

Thirty-five hundredths ____________ $\frac{3}{10}$ ____________

Seven millionths ____________ $\frac{19}{1000}$ ____________

Write each as a common fraction.

.9 = .37 = .691 = .08 =

Change each to a decimal fraction.

$\frac{4}{5}$ ____________ $\frac{9}{20}$ ____________ $\frac{89}{50}$ ____________

14. Add 3.12; 41.3; 68.94; and 39.77 ________
15. Subtract 23.9564 from 32.404 ________
16. Multiply .03 by .02 ________
17. Divide 160 by .04 ________
18. Find the cost of 7 pounds 7 ounces at $1.20 an ounce. ________
19. How much cloth, at $2.90 a yard, can be bought for $30. ________
20. What is the cost of 343 yards, 2 feet, 3 inches of tubing at $.16 per yard? ________

Score ________

Circulating Decimals

Name ______________________________ Date _______________

Principles

1. If any prime factors besides ________ and ________ are found in the denominator of a fraction in its lowest terms, the resulting decimal will never come to an end.

2. These decimals will contain a figure or set of figures which ______________________________

Vocabulary

3. A decimal in which one or more figures constantly repeat in the same order is called a ________________________________
____

4. The repeating set of figures is called a ______________.

Using Principle I, above, tell whether or not each of the following fractions will produce never-ending decimals.

$\frac{4}{9}$ __________ $\frac{1}{3}$ __________ $\frac{3}{25}$ __________

$\frac{1}{7}$ __________ $\frac{6}{50}$ __________ $\frac{17}{24}$ __________

Change each decimal fraction to a common fraction.

$.05 =$ $.2083 =$

13. Add $.20\dot{4}\dot{5}$, $.\dot{0}\dot{9}$, and .25. __________

14. Subtract $.0\dot{0}7\dot{4}$ from $.2\dot{6}$. __________

15. Divide $.\dot{7}\dot{5}$ by $.\dot{1}$. __________

Score __________

Compound Denominate Numbers

Name ______________________________ Date _______________

An example of a **compound denominate number** is 5 yards, 2 feet.

1. Which word refers to the 5 and 2? ___________________

2. Which word refers to the yard and feet measures?
____

3. Which word denotes that there are more than one measure?
____

---

4. Dollars, francs, and marks are measures of __________.

5. Drams, pounds, and ounces are measures of __________
________________________ .

6. Days, hours, and years are measures of ____________.

7. How many steps must a man take in walking from Kansas City to St. Louis, if the distance be 275 miles and each step is 2 feet 9 inches? __________

8. If it is now September 22 and Aaron's birthday is July 1, how many months and days must he wait for his birthday? __________

9. How many days and months is it until your next birthday? __________

10. What are these aliquot parts of 100?

$12\frac{1}{2}$ = $\qquad$ $33\frac{1}{3}$ = $\qquad$ 20 =

Score _____________

Ratio and Proportion

Name ______________________________ Date _______________

1. Ratio is the measure of the (relation, addition) of one number to another.
2. You can find a ratio by (dividing, adding) the numbers.

$$5 : 6 = \frac{5}{6}$$

3. Which number above is the ratio? ____________
4. Which number is the antecedent? ____________
5. Which number is the consequent? ____________

____

6. What is the ratio of a foot to a yard? ____________
7. What is the ratio of a yard to a foot? ____________
8. A proportion is a statement of two (equal, unequal) ratios.
9. Using numerals and ratio and proportion signs, write a statement that fifty has the same ratio to one hundred that one has to two. ____
10. Write a statement showing that the ratio of ten to one is the same as the ratio of one hundred to ten. ______________
11. The :: sign for proportion means the same as the (division, equal) sign.

	freezing point	boiling point
Fahrenheit Centigrade	32° 0°	212° 100°

12. How many degrees from freezing to boiling are there in Fahrenheit? ______________ In Centigrade? ______________
13. What is the ratio of a Farenheit degree to a Centigrade degree in its lowest term? ________________________________ What is the ratio of Centigrade to Fahrenheit?____________________________
14. Convert 70° Fahrenheit to Centigrade. ______________

Score ____________

Answer Key

INTRODUCTION. True or False: T, F, F, T. **Questions: 5)** formula. **6)** unit. **7)** an abstract. **8)** a concrete. **9)** a concrete. **10)** + − × ÷ . **11)** = : :: ( ) ________ .... ∴ .

NUMERATION. 1) period. **2)** billions. **3)** thousands. **4)** 2. **5)** 4. **6)** 8. **7)** 428,642. **8)** 500,060. **9)** 80,623,000. **10)** 800,000,000. **11)** IX, X, XI, XII. **12)** LXXXV, LXXXVI, LXXXVII, LXXXVIII, LXXXIX, XC. **13)** CCC, CCCC (or CD). D, DC, DCC, DCCC. **14)** DCCCC, CM.

ADDITION AND SUBTRACTION. 1) 3428. **2)** 18,312. **3)** 1738. **4)** 5839. **5)** $1557. **6)** $3200. **7)** 179. **8)** $1687. **9)** 2964. **10)** 7,914,100.

MULTIPLICATION AND DIVISION. 1) 15,292. **2)** 1,554,768.
3) 65,145,000.
4) Annex two zeros.
Multiply by 2 and subtract.

```
 641300
 -12826
-------
 628474
```

5) Multiply by 8.
Multiply the partial product by 4.

```
     4936
     3208
---------
    39488
  157952
---------
 15834688
```

6) 973. **7)** $22\frac{8}{17}$.
8) Use factors. For example, divide by 4, then divide that result by 6.

```
4)4524
  ----
6)1131
  ----
  188½
```

9) Use factors, as above. Or annex two zeros and divide by 4.

```
4)6847500
  -------
  1711875
```

10) 244,477.

PROPERTIES OF NUMBERS. Vocabulary: 1) integer. **2)** mixed number. **3)** prime number. **4)** composite number. **5)** multiple. **6)** divisor. **7)** composite number. **Prime factors: Of 45)** 3, 3, 5. **Of 72)** 2, 2, 2, 3, 3. **Of 98)** 2, 7, 7. **Greatest common divisor:** 30, 8, 14. **Least common multiple:** 36, 60, 210. **Problems:**

5639 5	875 2	78	6
+481 4	−749 2	×8	×8
6120 0	126 0	614 (3)	49 (3)

$$\begin{array}{r} 249 \\ 2\overline{)498} \end{array} \qquad \begin{array}{l} 6 \times 2 = 3 \\ 3 \end{array}$$

COMMON FRACTIONS. 1) 5. **2)** 9. **3)** proper. **Reduce:** $\frac{2}{9}$, $\frac{2}{3}$, $\frac{12}{17}$. **Change:** 33, 28, 75. **Add:** $\frac{29}{36}$, $10\frac{31}{84}$. **Subtract:** $\frac{16}{165}$, $\frac{23}{60}$. **Multiply:** $9\frac{3}{13}$, $4\frac{4}{5}$, 30 76. **Divide:** $\frac{3}{16}$, 9, $1\frac{1}{2}$, $\frac{9}{10}$.

DECIMAL FRACTIONS. Decimal fractions, left to right: .051, .07, .35, .3, .000007, .019. **Common fractions:** $\frac{9}{10}$, $\frac{37}{100}$, $\frac{691}{1000}$, $\frac{8}{1000}$. **Change to decimal fractions:** .8, .45, 1.78. **Problems: 14)** 153.13. **15)** 8.4476. **16)** .0006. **17)** 4000. **18)** $142.80. **19)** 11 yd. **20)** $55.

CIRCULATING DECIMALS. 1) 2, 5. **2)** repeat. **3)** circulate or circulating decimal. **4)** repetend. **Never-ending decimals, left to right:** yes, yes, no, yes, no yes. **Change:** $\frac{1}{18}$, $\frac{5}{24}$. **13)** $1.79\dot{6}$. **14)** $.\dot{2}5\dot{9}$. **15)** $6.\dot{8}\dot{1}$.

COMPOUND DENOMINATE NUMBERS. 1) number. **2)** denominate. **3)** compound. **4)** value. **5)** weight. **6)** time. **7)** 528,000. **8)** 9 mo. 9 da. **9)** **10)** $\frac{1}{8}$, $\frac{1}{3}$, $\frac{1}{5}$.

RATIO AND PROPORTION. 1) relation. **2)** dividing. **3)** $\frac{5}{6}$ **4)** 5. **5)** 6. **6)** $\frac{1}{3}$. **7)** 3. **8)** equal. **9)** 50 : 100 :: 1 : 2. **10)** 10 : 1 :: 100 : 10. **11)** equal. **12)** 180, 100. **13)** $\frac{9}{5}$ or $1\frac{4}{5}$, $\frac{5}{9}$. **14)** 126°.

Teaching Eighth Grade

Using *Ray's New Higher Arithmetic*

Teaching Eighth Grade

While teaching eighth graders you will probably feel a special responsibility to see that the students are proficient in all the arithmetic they are supposed to have learned thus far. During this year when you spot weaknesses, do some reteaching and strengthen the skills as needed.

From the textbook, your students can begin by studying percentage. Then they can practice those skills in many consumer and business applications in the chapters which follow the percentage chapter. Supplement the textbook by using many modern examples. Study newspaper ads about savings, installment buying, and other financial matters. Make up problems. Find out, for instance, how much is actually paid for a car when it is bought with an installment loan. Find out which savings plan brings the best return on money. Solving problems with percentages is a skill everyone needs in our modern society.

Students who have good arithmetic skills can study the other topics in the textbook, such as alligation and series. Some few students may study algebra, probabilities, navigation, computers, or other topics from other books.

A planning guide is given below for a basic eighth grade course and another for an advanced eighth grade course. Notice which chapters are covered in each. Then set a goal for your students to cover at least the basic course and as much of the advanced as you think they can manage. Help your students to work hard and learn a lot this year.

Basic Course

(Ray's Higher Arithmetic)

Study chapters XIV and XV and the topics you consider most useful in chapter XVI. Then study chapter XXII.

Remember to supplement the work with modern examples from newspapers and from other real-life sources.

Advanced Course

(Ray's Higher Arithmetic)

Study chapters XIV through XXIII. Supplement with modern life examples, especially in the business and consumer topics.

The following pages show ready-made tests for each eighth grade unit. You may make copies of these tests for use with your students.

Percentage

Name ______________________________ Date _______________

1. Percentage refers to problems based on (100, Arabic numerals).

5% of 60 is 12

2. In the example above, which number is the base?

____

3. Which number is the rate? _____________

4. Which number is the percentage? _____________

---

5. Find $16\frac{2}{3}$% of 1932. _______________

6. Find 35% of $\frac{2}{7}$. _______________

7. 750 is what per cent of 12,000? _____________

8. $.15 is what per cent of $2? _____________

9. $3.80 is 5% of what amount? _____________

10. 192 is $\frac{6}{10}$% of what number? ___________________

11. 96 is 100% more than what number? _____________

12. $480 is $33\frac{1}{3}$% more than what amount? _______________

13. If you sleep 8 hours each night, what per cent of the time do you spend sleeping? _____________

14. A man owned 50% of a business. To raise some money he sold 25% of his share. How much of the business did he still own?

____

Score _____________

Percentage Without Time

Name ______________________________ Date _______________

1. Mr. Williams invested $1450 and made $14\frac{1}{2}$ per cent profit. What is the amount of his profit? ___________________

2. If you buy something for $80 and sell it for $90, what per cent would you gain? _______________

3. A rancher lost $6\frac{1}{6}\%$, or 40, of his sheep during winter storms. How many sheep were in his original flock?

____

4. If you want to sell candy for $5 a box and make at least $33\frac{1}{3}\%$ profit, what is the most you can pay for each box?

____

5. Mr. Henry owned 250 shares of stock. The company declared a dividend of 2% payable in stock. How many more shares did Mr. Henry receive? _______________

6. Mr. James paid $35 for 100 shares of stock and he received a dividend check for $280. What per cent did he make on his money? ________________

7. Since Mr. James bought his stock, it has gone up and the market value was $50 when the dividend was declared. What per cent did the company pay on its stock? _________________

8. A $1000 bond paying 9% interest now sells at a 25% discount. If you buy that bond, what must you pay for it and what per cent interest will you make on your money? _____________

____

9. An architect charges $2\frac{1}{2}\%$ for his services. What is his fee on a building costing $950,000? _________________

10. A real estate agent received $3\frac{1}{2}\%$ commission when he sold a house for $80,000. What was the amount of his commission?

____

11. Mr. Donn bought 100 shares of stock with a market value of $22.50 per share and with 3% brokerage commission. How much did he pay? ______________________

Score _____________

Percentage With Time, and Partnership

Name ______________________________ Date _______________

Vocabulary

interest **principal** **rate of interest** **amount**

1. Which term refers to money that is charged for using money? ______________________________

2. Which refers to the total of the money borrowed and the money charged for interest? ______________________________

3. Which refers to the amount of money borrowed? ______________________________

4. Which tells the per cent that is charged on the principal? ______________________________

---

5. What interest rate must you earn to double your money in 10 years? _______________

6. To double your money in 5 years you need a (higher, lower) rate of interest than above.

7. What principal will produce $45 a month at 9%? ______________________________

8. What is the difference between the annual and the compound interest of $5000 for 6 years at a rate of 6% per year? ______________________________

9. Sarah and Jane had a summer business to which Sarah contributed $25 start-up costs and Jane contributed $30. Their profits were $385. How much did each girl make? Sarah _______________, Jane _______________.

10. Jim and Nate had $210 at the end of the summer in the treasury of their business. But Jim had contributed $300 to start and Nate had contributed $400. How much did each boy lose? Jim _______________, Nate _______________.

Score ___________

Alligation

Name ______________________________ Date ______________

1. Find the average price of a commodity which was bought at these prices.

6 lb. at 80 cents per lb.
15 lb. at 50 cents per lb.
5 lb. at 60 cents per lb.
9 lb. at 40 cents per lb.

____

2. A student's arithmetic test grades were 95, 80, 90, 80, 80, and 85 per cent. What was his average? ________________

3. One kind of sugar costs 18 cents a pound and another kind costs 10 cents a pound. A bakery wishes to use sugar at an average of 12 cents a pound. What relative amounts of the two sugars must they use? ____

4. What relative amounts of silver $\frac{3}{4}$ pure and silver $\frac{9}{10}$ pure will achieve a mixture $\frac{7}{8}$ pure? ________________________

5. A man bought 80 shares of stock in a food company for $72 a share. When the stock declined to $50 a share, he wanted to purchase more so as to make his average price $60. How many shares must the man buy? ____________________

Score ____________

Involution and Evolution

Name ______________________ Date ______________

1. Find the second power of 81. ______________

2. Find the third power of $16\frac{1}{10}$. ______________

3. Find the square of 309. ______________

4. Find the square of $\frac{10}{21}$. ______________

5. Find the cube of 36. ______________

6. Find the fourth power of 91. ______________

7. Find the fifth power of 11.1. ______________

8. Find the square root of 625. ______________

9. Find, by factoring, the square root of 1225.

____

10. Find the cube root of 13824. ______________

Score ____________

Series

Name ______________________________ Date ______________

1. What is the next term of the series 3, 7, 11? __________
What is the twelfth term of the series? ________________

2. Find the eighth term of the series 100, 96, and so forth.
____

3. Find the tenth term of the series .025, .037, and so forth.
____

4. There are 20 terms in a series from 4 to 80. What is the sum of the terms? ________________

5. If you count by sevens beginning at 450 and ending ten terms later, what is the sum of the terms you count? ____________

6. If you count from 8 to 54 with only one term between them, what will that term be? ________________

7. If you count from 4 to 40 with only 4 terms all together, what are the two missing terms? __________________________

8. A series begins with 2 and the ratio is 2. That is, each term is twice the one before it. What is the tenth term? __________

9. The first term is 1 and the ratio is 3. What is the seventh term? ________________

10. The first term is 64 and the ratio is ½. What is the twelfth term? ________________

Score ____________

Mensuration

Name ______________________ Date ____________

Vocabulary

straight **curved** **horizontal** **vertical**

1. The word **line** used alone means a ______________ line.
2. A line parallel with the horizon is a ______________ line.
3. A line running opposite to the horizon is a ______________ line.

---

angle **surface** **solid** **area** **volume**

4. Has length and width, but not thickness: ____________
5. Has length, width, and thickness: ____________
6. A defined amount of space on a surface: ____________
7. Space which is measured by cubic units: ____________

Draw an acute triangle. An obtuse triangle. A right triangle.

11. Find the area of a rectangle which is 9 feet 4 inches by 2 feet 5 inches. ______________________
12. Find the area of a circle with a diameter of 10 feet. ______________________
13. Find the surface of a rectangular prism 4 feet high, standing on a base 2 feet square. ______________________
14. Find the volume of a cylinder 20 inches high whose circumference is 39.27 inches. ______________________
15. Find the volume of a pyramid 48 feet high, on a base 18 inches square. ______________________

Score ____________

Answer Key

PERCENTAGE. 1) 100. **2)** 60. **3)** 5%. **4)** 12. **5)** 322. **6)** $\frac{1}{10}$. **7)** $6\frac{1}{4}$%. **8)** $7\frac{1}{2}$. **9)** $76. **10)** 32,000. **11)** 48. **12)** $360. **13)** $33\frac{1}{3}$%. **14)** $37\frac{1}{2}$%.

PERCENTAGE WITHOUT TIME. 1) $210.25. **2)** $12\frac{1}{2}$%. **3)** 920. **4)** $375. **5)** 5. **6)** 8%. **7)** $5\frac{3}{5}$%. **8)** $750, 12%. **9)** $23,750. **10)** $2800. **11)** $2317.50.

PERCENTAGE WITH TIME AND PARTNERSHIP. Vocabulary: 1) interest. **2)** amount. **3)** principal. **4)** rate of interest. **Problems: 5)** 10%. **6)** higher. **7)** $6000. **8)** $22.596 or $22.60. **9)** Sarah $175, Jane $210. **10)** Jim $210, Nate $280.

ALLIGATION. 1) 54¢ per lb. **2)** 85. **3)** 1 : 3, or 1 lb. of 18¢ sugar to every 3 lb. or 10¢ sugar. **4)** 1 : 5, or 1 lb. of $\frac{3}{4}$ silver to 5 lb. of $\frac{9}{10}$ silver. **5)** 96.

INVOLUTION AND EVOLUTION. 1) 6561. **2)** 4173.281. **3)** 95,481. **4)** $\frac{100}{441}$. **5)** 46,656. **6)** 68, 574,961. **7)** 16,850,581,551. **8)** 25. **9)** 35. **10)** 24.

SERIES. 1) 15, 47. **2)** 72. **3)** .133. **4)** 840. **5)** 5335. **6)** 31. **7)** 16, 28. **8)** 256. **9)** 729. **10)** $\frac{1}{32}$.

MENSURATION. Vocabulary: 1) straight. **2)** horizontal. **3)** vertical. **4)** surface. **5)** solid. **6)** area. **7)** volume. **Draw:** Acute triangle with all angles less than 90°. Obtuse triangle with one angle more than 90°. Right traingle with one angle 90°. **Problems: 11)** 22 sq. ft. 80 sq. in. **12)** 78.54 sq. ft. **13)** 40 sq. ft. **14)** 1 cu. ft. **15)** 36 cu. ft.

Projects and Games

Projects

Arithmetic All Around You

(all grades)

From time to time bring to class a number item you used in everyday life, and encourage your children to do the same. Some possibilities are: menu, cash register tape, airline or bus schedule, newspaper bill, electricity bill. Help your children read and understand these items.

Sometimes post such an item on the bulletin board. Make up a problem that can be answered by studying the item, and post it too. Have the children solve the problem during any spare moments they have, but they are to keep the answer secret from other pupils. Then near the end of the day see how many have the correct answer.

Counting Rhymes

(lower grades)

1, 2,
Buckle my shoe;

3, 4,
Knock at the door;

5, 6,
Pick up sticks;

7, 8,
Lay them straight;

9, 10,
A big fat hen;

11, 12,
Dig and delve;

13, 14,
Maids a-courting;

15, 16,
Maids in the kitchen;

17, 18,
Maids in waiting;

19, 20,
My plate's empty.

One, two, three, four,
Mary at the cottage door,
Five, six, seven, eight,
Eating cherries off a plate.

One, two, three, four, five,
Once I caught a fish alive,
Six, seven, eight, nine, ten,
Then I let it go again.
Why did you let it go?
Because it bit my finger so.
Which finger did it bite?
The little finger on the right.

Thirty days hath September,
April, June, and November;
All the rest have thirty-one,
Excepting February alone,
And that has twenty-eight days clear
And twenty-nine in each leap year.

One's none,
Two's some,
Three's many,
Four's a penny,
Five's a little hundred.

X shall stand for playmates Ten;
V for Five stout stalwart men;
I for One, as I'm alive;
C for Hundred, and D for Five;
M for a Thousand soldiers true,
And L for Fifty, I'll tell you.

Find

(all grades)

Find numbers in newspapers. What do they tell you? Can you make up problems from the information?

Geometric Art
(middle and upper grades)

1) Use triangles, rectangles, or other congruent figures and arrange in a pattern, such as floor tile patterns. Congruent figures are the same in shape and size, but may be different in color. Children may discover that various rectangles and parallelograms may be arranged in row-like patterns, but for continuous, "over-all" patterns only squares, equilateral triangles, and regular hexagons will work. Can they figure out why? (All angles joining at one point in a design must total 360°, a full circle. By experimenting or by figuring, children can find that only angles of 60°, 90°, or 120° can be arranged to total 360°.)

2) Fold a piece of paper and cut out a symmetrical design. Fold the paper in one direction, then in the opposite direction and cut out a design that has two lines of symmetry.

3) Use compasses and draw symmetrical designs. Draw non-symmetrical designs. Color the spaces in the designs.

4) Use paper and construct cubes, cylinders, cones, and other three-dimensional shapes.

5) Place a grid of quarter-inch squares over a drawing that you want to enlarge. The squares may be drawn directly over the picture or on a tissue to be placed over the picture. Then carefully redraw each square onto squares one-inch in size. This makes the picture sixteen times larger, and you don't have to be an artist to do it.

Recipe
(lower and middle grades)

Follow a recipe together. Read, measure, count, mix. Then cook, if needed. Count again when cutting or passing out the treats to the whole group.

Shopping
(lower and middle grades)

Use a large grocery ad or other ad from a newspaper. Children take turns making up problems for others in the class to work. One type of problem is to name two or three items the child wants to buy and ask what they will cost all together. Another type of problem is to name a total amount of money the child can spend, and ask what he can get with it. Problems of this type have a variety of answers.

Store

(lower grades)

Set up a grocery store with empty Jello boxes, cereal boxes, cans, and so forth. Mark prices for everything and use play money to play Store.

Variations. Set up specialty stores—hardware, book, popcorn, or other. Sometimes use real money. For instance, used books can be donated to the store and class members may buy them. Use the proceeds to buy books for the classroom library.

Telephone

(lower grades)

Learn to call a parent at work, a neighbor who will help in emergencies, a friend. Learn to use both dial and touch-style phones.

Games to Learn Numbers and Counting

Exercises

A child demonstrates an exercise—toe touch, high kick, or other—and calls out a number, for instance 7. Then he leads the group in doing the exercise seven times. Have many turns for plenty of exercise, as well as plenty of number practice.

Guess My Number

Two players each begin with twenty pieces of play or real money, all of the same denomination. For example, use all nickels. One player writes a secret number from 1 to 100. The other player guesses and pays one coin for his guess. The first player points upward if the secret number is higher than the guess, and he points downward if it is lower. The second player continues guessing and paying until he guesses correctly. Then he writes a secret number and the first player pays him to guess.

If one player runs out of money, the other is the winner. If neither player runs out of money, they can stop after an even number of turns and count to see who has the most money.

Variations. For easier games adjust the range of numbers. For instance, guess numbers from 1 to 10, and use only five pieces of money.

Number Permutations

Two or three children may play. Each player makes a set of ten cards, with the digits from 0 to 9. Players mix up their cards and place their own piles face-down in front of themselves.

When play begins, each player draws three of his cards and makes the greatest number he can with them. The player with the greatest number wins a point. The player with the most points by the end of a specified time period is the winner.

Variations. Use only one pile of cards, and have the children first draw three blanks on their papers, like this:

____ ____ ____

Then someone draws the cards one at a time. As each number is drawn the players write it on their blanks. After a number is written its position cannot be changed. The player who ends with the largest number wins.

To practice with numbers in the thousands, the children may draw four cards.

Pirate Ship

Two players each prepare two graphs as shown below. Each places two pirate ships, each in three consecutive squares, which can be vertical, horizontal, or diagonal. Players should not let opponents see their home-waters graphs.

The object is to sink the opponent's ships. The first player calls out a shot in this manner: A-5, or D-1, and he marks the guess on his graph of enemy waters, and that particular square should not be fired on again. His opponent tells him whether the shot is a hit or not. The second player then fires. It takes two hits to sink a ship. Play continues until one pirate loses both ships.

Variations. This game gives practice in graph reading. The graph can be enlarged and more ships used.

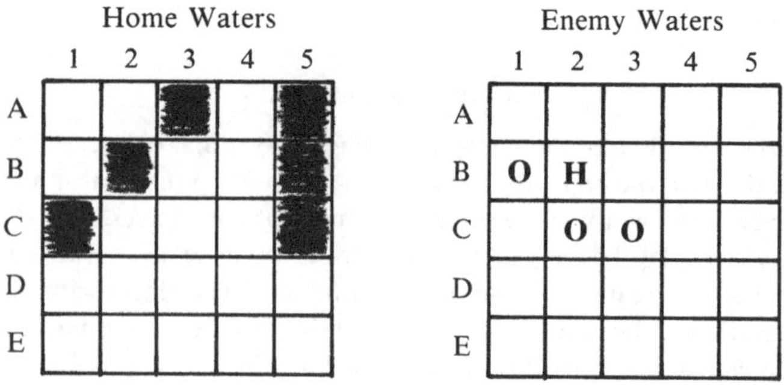

Throwing and Catching

Children throw balls or beanbags to each other in pairs or across a circle. Players count the catches. When a child misses, the count must begin at 1 again. How high can the group or pair count without a miss?

Zip

Players sit in a circle. One begins counting with 1, the next says 2, and so on around the circle. But players must say "zip" for 5 and all multiples of 5. A player who hesitates or forgets must stand. He remains standing until another turn comes around, when he may count correctly and sit down again.

Variations. When 5's are easy for the group, try 2s, 7s, or any number. With older children, try two numbers at a time, for instance saying "zip" for multiples of 5 and "zap" for multiples of 7.

Another variation is to add rhythm with a clapping pattern, which can be simple or complex, according to the age of the children. For instance, a four-count pattern might be: 1) clap hands on lap, 2) clap hands together, 3) snap fingers, 4) say number.

A lead-up game for beginners is for all to count together and substitute a clap for the word "zip."

Games to Learn Number Facts

Concentration

Make eight pairs of cards with a problem on one and its answer on the other. These can fit most any topic your children are studying. Mix the cards and lay them face down in four rows of four.

The first player turns over two cards. If they match he keeps them. If they do not, he turns them down again. The second player then takes his turn. When all pairs are matched, the one with the most cards is the winner.

Variation. With some sets of cards, add the rule that the children must turn over one card, name the match they are looking for, and then turn over a second card.

Favorite Race

Here is a use for old game boards that have some kind of track on them. Call it a ski race or auto race or whatever your children are interested in. Have problems written one to a card. Each player or team solves a problem in turn and moves their marker forward an agreed number of spaces if the answer is correct.

Variation. Use cards of basic facts without the answers. Each player draws a card and moves his marker forward as many spaces as the answer. He collects a scoring counter or M&M each time he passes GO.

Good-Guy

Make pairs of matching cards of any facts your children are learning. For instance, a card may read $6 + 8$ and its matching card read $8 + 6$. Or a card may read 7×3 and its matching card read $(3 + 4) \times 3$. Add two extra cards with a comic strip character on each. Find a "good guy" and a "bad guy" for these cards.

Have three or four players. Deal out all the cards. The player with the good guy lays it down and takes the first turn. He draws a card from the player to his left. Then he lays down any matches he has. Each player in turn draws and lays down his pairs. The game is over when only the bad-guy card is left. The winner is the one with the most pairs who does not have the bad-guy card.

Multiply or Divide

Use a set of flashcards which do not have answers. Mix them and place the pile face-down. Each player draws a card and tells the answer. The player with the largest product or quotient wins a point. Use tokens or M&Ms for the points.

Number, Please

Whisper to each pupil a number, being sure that at least two children have each number given out. Use numbers according to the facts your class is learning. For instance, if they are learning the multiplication table of 7s, give out 7, 14, 21, and so forth.

The leader calls out a problem, such as 5×7. Students with the number 35 stand as quickly as they can. The leader asks the first one standing, "Number, please?" He answers, "35," and sits down. The leader calls another problem. Keep the game moving as rapidly as possible and play for only six or eight minutes.

Sassy Sixes

Each player writes down six of the sums, or remainders, or products, or quotients the class is learning. A leader draws one card from a flashcard pile. Anyone who has that sum on his paper crosses it off. The first player to cross off all his numbers is the winner.

The Greatest

Play in groups of four or five. Use a pile of flashcards with problems only, no answers. Each player draws a card. The player holding the greatest sum (or difference, or product, or quotient) wins all the cards for that round. Draw again and continue until the pile is gone. The winner is the player with the most cards.

Games For Solving Problems

Review Relay

Put a copy of the textbook at the chalkboard in front of each team and open the books to the page being reviewed. Or write the problems on cards. The first player on each team goes to the board, writes the problem, works it, and sits down. The next player does the next problem, and so forth. But if a player makes a mistake, the following player must correct it before doing his problem.

Variation. Have Team A start with the first problem and Team B start with the last problem and work backward. This way it is a mystery who is ahead until one team completes its problems.

Subtract or Add

Write numbers from 0 to 99 on cards. Mix them and place the pile face-down. Each player draws two cards and subtracts the smaller number from the larger. The player with the largest difference wins a point. Use checkers or other tokens to keep track of points. Or use M&Ms or other edible prizes, in which case the children may prefer to eat them as they go instead of saving them to count up the total.

For addition practice, write numbers to 50.

Tic-Tac-Toe Squares

Two people may play. The first player draws a tic-tac-toe grid and places a number in the lower right corner. This number must end up being the sum of the column above it and the sum of the row to the left of it. Those numbers, in turn, must be the sums of their respective rows or columns.

Each player takes turns placing a number in the grid. If a player writes a number which will not work, he loses. If no one makes an error, the first player loses because his beginning sum did not stump his opponent.

Example of completed grid:

6	2	8
4	3	7
10	5	15

Example of an error:
(Six was the last number placed, and it will not work because the sum of that column must be 5.)

	6	
7		12

Variations. This game can also be played with subtraction. The first player puts the remainder in the lower right square.

For a class version of this game, use teams. Place the grid on the chalkboard and let players from each team take turns coming up to write a number on the grid.

For a puzzle version of this game, have the children see if they can correctly fill in a sixteen-square grid.

Twenty

For two players you need forty-five index cards. Put the number 1 on five cards, 2 on five cards, and so on up to the number 9. An alternate way to prepare the cards is to use domino-like dots instead of figures.

Mix the cards and place them face-down in a pile between the players. Each player draws two cards.

The object of the game is to play cards face-up, one at a time, until the numbers total 20. The first player plays a card face-up in front of him, then draws another card. The second player does the same. When a player holds only cards that would make his total more than 20, he must use his turn to discard and draw another. The discard pile is face-up beside the first pile. Turns continue, either playing or discarding, until one player makes a total of exactly 20.

Games of Geometry and Measures

Treasure Hunt

Give each team a list of things to find in the room. The list can vary according to the topics the class is studying. Some examples are: an object 16 inches long, something that weighs more than 1 pound but less than $1\frac{1}{2}$ pounds, something with an acute angle, a polygon. The team which finds all its items first, or finds the most items within a time limit, is the winner.

Walk the Tightrope

Have the children measure two lines several feet or yards or a rod long, depending on the measures they are learning. Stretch string along the lines. Then have a tightrope relay.

Each player, in relay fashion, must walk the length of the string and back without stepping off it. A player who steps off, must start at the beginning again. The first team to finish successfully wins the game.

Make this more difficult by requiring tightrope walkers to balance an object, such as a paper-towel tube, on the palm of one hand. Or they may balance a book on their heads.